SCIENTIFIC AMERICAN EXPLORES BIG I[illegible]

Extreme Animals

The Editors of *Scientific American*

SCIENTIFIC AMERICAN EDUCATIONAL PUBLISHING

New York

Published in 2024 by Scientific American Educational Publishing
in association with **The Rosen Publishing Group**
2544 Clinton Street, Buffalo NY 14224

Contains material from Scientific American®, a division of Springer Nature America, Inc., reprinted by permission, as well as original material from The Rosen Publishing Group®.

First Edition

Scientific American
Lisa Pallatroni: Project Editor

Rosen Publishing
Erica Grove: Compiling Editor
Michael Moy: Senior Graphic Designer

Cataloging-in-Publication Data
Names: Scientific American, Inc.
Title: Extreme animals / edited by the Scientific American Editors.
Description: First Edition. | New York : Scientific American Educational Publishing, 2024. | Series: Scientific American explores big ideas | Includes bibliographical references and index.
Identifiers: ISBN 9781725350397 (pbk.) | ISBN 9781725350403 (library bound) | ISBN 9781725350410 (ebook)
Subjects: LCSH: Animal behavior–Juvenile literature. | Animals–Juvenile literature. | Animals–Miscellanea–Juvenile literature.
Classification: LCC QL751.5 E987 2024 | DDC 591.5–dc23

Manufactured in the United States of America
Websites listed were live at the time of publication.

Cover: reptiles4all/Shutterstock.com

CPSIA Compliance Information: Batch # CWSA24.
For Further Information contact Rosen Publishing at 1-800-237-9932.

CONTENTS

INTRODUCTION

When we think of the extreme characteristics and behaviors of animals, we tend to focus on a few in particular. We might think of the cheetah, the fastest land animal, or the blue whale, the biggest animal. But there are many other species with characteristics that seem extraordinary to humans, and many that exhibit behaviors we would consider strange—even among animals we tend to take for granted. For instance, the noble false widow spider—which measures less than an inch (7 to 14 mm)—is able to kill prey much larger than itself, including rodents, lizards, and small bats. Birds use photoreceptors in their brains to track daylight length, which trigger activities in their bodies that help them determine when to migrate. Earthworms are able to regenerate heads and tails, and sometimes they will even end up regrowing two heads or tails in a process known as bifurcation. This volume explains the traits that make a wide range of animals unique, from insects to blue whales and everything in between.

Section 1, "Wild Behavior," explores strange and astonishing activities animals engage in. Section 2, "Amazing Animal Anatomy," looks at unexpected physical characteristics animals possess and the ways in which these impact their behaviors. Section 3, "Animals in Extreme Places," explains some of the weird and seemingly inhospitable habitats animals call home. Section 4, "Strange Connections Between Humans and Animals," explores the unique ways certain types of animals relate to humans and surprising instances in which humans and animals share behaviors and characteristics. Finally, Section 5, "Extreme Animal Evolution," looks at how animals have evolved over time to become the unique creatures they are today.

Section 1: Wild Behavior

50,000 Worms Tangled Up in a Ball Unravel in an Explosive Burst when a Predator Appears

By Jack Tamisiea

Anyone who's grappled with jumbled headphones knows the difficulty of disentangling snarled cords. A tight knot is nothing for a California blackworm, however. These tiny worms twist together by the thousands to form tightly packed blobs reminiscent of a forkful of squirming spaghetti. While these tangles take minutes to form, intertwined blackworms can wriggle free in a matter of milliseconds.

Now scientists have finally straightened out how these legless escape artists use only a simple collection of muscles and neurons to seamlessly slither out of tight tangles. "We thought that if worms can solve this untangling problem, so could we," says Vishal Patil, an applied mathematician at Stanford University. In a study published today in the journal *Science*, Patil and his colleagues used mathematical simulations to pinpoint the movements blackworms use to rapidly untangle themselves.

At just a couple of centimeters long, California blackworms (*Lumbriculus variegatus*) are easy to overlook. Yet these aquatic worms, which are common grub for aquarium fish, exemplify strength in numbers. To conserve their moisture or maintain their temperature, anywhere from five to 50,000 blackworms intermingle to form writhing blobs seemingly right out of a creature feature. While these tangles are tight, the earliest sign of a predatory diving beetle will send the fleshy worms squirming in all directions.

Harry Tuazon, now a bioengineering Ph.D. student at the Georgia Institute of Technology, caught a glimpse of this nearly instantaneous response when he observed blackworms wriggling across a petri dish in the lab. "I pointed a UV light toward a ball of worms, and it suddenly exploded," he says. "It was mesmerizing."

Tuazon became enamored with these worm balls, which unraveled in a few tens of milliseconds—a fraction of an eyeblink. He took microscope videos of the movements of individual worms before gradually adding more of the animals to the mix to increase the complexity of the balls. To untangle the physics behind these worm webs, he teamed up with Patil, then a Ph.D. student at the Massachusetts Institute of Technology specializing in the geometry of knots and other complex systems.

Patil noticed something interesting when he watched microscope videos that Tuazon had shot of a single worm responding to a small electric shock. As the worm reacted to the stimuli, it moved its head clockwise before reversing course and turning counterclockwise in a repeating motion. This created a figure-eight pattern known as an alternating helical wave.

To create accurate mathematical models of the worms' helical gait, Patil also needed to see how the individual worms moved inside the tangles. This proved difficult for Tuazon because the worm balls were surprisingly impenetrable. The worms were submerged in water, so x-rays proved unsuccessful. Microcomputed tomography scans only offered low-resolution glimpses. Finally, Tuazon determined that the team's best bet was sound. He used an ultrasound machine to create images of a cluster of living worms that were placed inside gelatin.

The ultrasound images allowed the researchers to plot more than 46,000 points of data on the movements of the individual worms, who were constantly in contact with other worms inside the writhing jumble. Patil and his colleagues created mathematical models of the worms' movements and ran them through three-dimensional simulations of the tangles.

The researchers found that employing the alternating helical gait motion allowed the worms to seamlessly untangle themselves. They also found that a slightly altered movement, where the invertebrates corkscrewed in mostly one direction, helped create the tangles. "If it spends longer winding in one direction before switching to winding in the other direction, you get the tangling behavior," Patil says. "If the worm rapidly switches between winding clockwise and anticlockwise, you get untangling behavior."

According to Eleni Panagiotou, a mathematician at Arizona State University, who studies the effects of entanglements in physical systems, the versatility of this movement allows the worms to fine-tune their tangles. "California blackworms [exhibit] this optimal threshold where they can both tightly entangle and disentangle quickly," says Panagiotou, who was not involved in the new study but wrote a related perspective piece in *Science*. If they twist too much, the tangle becomes too tight, and the worms lose their ability to escape. If they don't twist enough, the protective tangle won't form.

She thinks similar motions can help untangle an array of complicated knots, which are ubiquitous in both nature and the artificial world. For example, tangled filaments naturally occur in everything from gnarled root networks to tightly wound DNA strands. Humans have long employed tangles to fashion rope and weave fabric. "The researchers not only explain what nature is doing in these worms, but they also create a map of the possibilities in other systems and contexts," Panagiotou says.

The team thinks these worms' movements can possibly help researchers program stringlike filaments in soft robotic systems to actively shape-shift. This could create flexible bandages that change shape as wounds heal or water filters that can be fine-tuned to sift out certain particles.

Patil thinks the proof is in the worm blob. "These worms give us the general principles behind tangling and untangling and potentially a toolbox for manipulating other systems," he says. "It's not just a mathematical model because we know that worms can do it."

About the Author

Jack Tamisiea is a science journalist based in Washington, D.C., who covers natural history and the environment. Follow Tamisiea on Twitter @jack_tamisiea.

Tiny Spider Fells Prey Many Times Its Size

By Jack Tamisiea

Dawn Sturgess looked out her bedroom window in southern England one day last summer just in time to witness an arachnophobe's nightmare: a shrew was tangled in a spider's web, and its ability to struggle was quickly weakening as the arachnid's venom shut down the tiny mammal's nervous system.

Over the next 20 minutes, Sturgess watched the female noble false widow spider hoist the immobilized, matchbox-sized Eurasian pygmy shrew—which was many times larger than the spider—nearly 10 inches upward into the rafters of the home, where the tiny predator chemically liquified and consumed its meal.

"I stood on the other side of the window, watching, for ages," says Sturgess, a citizen scientist who filmed the entire interaction on her phone. "I was fascinated by the spider's strength and determination to take its prey home!"

While this is the first confirmed case of a noble false widow spider devouring a shrew in the United Kingdom, it isn't the first time these ravenous arachnids have been seen taking down relatively large prey. According to a recent *Ecosphere* study co-authored by Sturgess and researchers at the Venom Systems & Proteomics Lab at the University of Galway in Ireland, this is the third documented case of a noble false widow spider preying on a vertebrate in the past five years. This suggests that small mammals such as the pygmy shrew—a protected animal in the U.K.—may be a more frequent item on spiders' menus than previously thought.

Noble false widow spiders (*Steatoda nobilis*) are native to the Canary Islands. They reached England in the 1870s, after stowing away on ships carrying bananas and ornamental plants. The spiders have since spread into mainland Europe and the Mediterranean and eventually into South America and California. They adapt well to unfamiliar regions because they are cold-tolerant and relatively long-lived—and can pump out up to 1,000 offspring in a single year.

Another helpful factor in their longevity is their versatile diet. John Dunbar, one of the University of Galway researchers who led the new study, says noble false widow spiders have been known to feast on a vast range of arthropods, from flies and wasps to centipedes and other spiders. The new finding adds evidence that these spiders—whose length maxes out at half an inch—also hunt bigger game. In recent years, Dunbar's team has also seen a common ground lizard and a small bat tangled up in a noble false widow spider's sticky web.

"The fact that this spider was able to lift a shrew 10 times its size was truly remarkable," says Jose Valdez, an ecologist at the German Center for Integrative Biodiversity Research, who published a literature review of vertebrate-eating arthropods in 2020. "The ability to lift such a large prey item suggests that the spider may be well adapted to hunt larger prey in its environment," adds Valdez, who was not involved in the new study.

Many vertebrate-eating spiders are mammoth, at least in arthropod terms. A prime example is the aptly named goliath bird-eating spider, the world's largest spider. With leg spans up to 11 inches across, this tarantula can turn just about anything smaller than itself into dinner, including frogs, lizards, mice and, of course, birds.

Unlike goliath birdeaters, the much smaller noble false widow spider can't outmuscle its prey. Instead it relies on a deadly combination of strong silk and powerful venom. Its web, a messy tangle of silk threads, makes an excellent snare. The spider then injects a potent concoction of neurotoxins that can paralyze small vertebrates such as pygmy shrews in a matter of minutes, Dunbar says.

Then it's time to eat. The venom is loaded with proteins that break down tissues, and the spider also expels digestive juices to help liquify the meal—and then it slurps up the soupy remains, sometimes over the course of several days. In the case of the pygmy shrew, the only scraps the researchers found after three days were inedible fur, bone and skin.

Among noble false widow spiders, so far only adult females have been observed eating vertebrates. This is likely because female spiders need to gather as much food-derived energy as possible to produce their prodigious numbers of offspring.

"A vertebrate would be a fine meal for a spider, so when they have the chance, they pounce," says Whit Gibbons, a herpetologist at the University of Georgia, who was not involved in the new paper. Last year Gibbons published a study that found that around 30 percent of spider families consume vertebrates. For example, a triangulate cobweb spider (which comes from Eurasia and is a close relative of the noble false widow spider) has been observed subduing a garter snake that was 355 times heavier than itself. Similarly sized brown widow spiders have successfully dispatched rats. "I think spiders in general fall into one of the many categories of animals that engage [with] a broader diversity of prey than assumed," Gibbons says.

Despite noble false widow spiders' taste for large prey, Dunbar doubts the arachnids will ever pose a serious threat to vertebrates such as pygmy shrews. But he thinks it is important to keep tabs on what these eight-legged invaders are eating. "It is very likely we will see other types of small vertebrates falling prey to the noble false widow [spiders] as they continue to expand their range," he says. "It adds a new dimension to their invasiveness."

About the Author

Jack Tamisiea is a science journalist based in Washington, D.C., who covers natural history and the environment. Follow Tamisiea on Twitter @jack_tamisiea.

Social Animals Seek Power in Surprisingly Complex Ways

By Lee Alan Dugatkin

Power in nonhuman animals used to play out in such a tidy and simple way. Bigger, stronger animals beat up smaller, weaker ones. The vanquished slunk away, and the victor claimed the prize. Or so we thought.

To be sure, there are enough of these sorts of brutish battles going on in nature to make war-of-all-against-all theorist Thomas Hobbes smirk. But we now know the quest for power in the animal kingdom is oh-so-much more subtle, interesting and, dare I say, beautiful than we previously thought.

The pursuit of power—which I define here as the ability to direct, control or influence the behavior of others and/or the ability to control access to resources—affects nearly every aspect of the lives of group-living animals. Winners in the struggle for power get more food, or more mates, or better, safer living quarters, and sometimes they get a combination of such spoils.

The strategic aspects of power in animals are mind-boggling. We biologists used to think animals were like simple robots that responded to fixed algorithms determined solely by their genes. During the breeding season, male stickleback fish become intensively territorial and take on a bright red coloration on their underside that attracts females. Dutch biologist Niko Tinbergen, who shared a Nobel Prize for founding the field of animal behavior, discovered that if you show a stickleback territory-holder virtually anything that's red, it attacks: "Even a red mail van passing our windows at a distance of 100 yards," Tinbergen wrote, "could make the males in the tank charge its glass side in that direction." In time, though, spurred on by evolutionary biologists John Maynard Smith and George Price's paper using game theory to analyze nonhuman behavior, ethologists realized that how one animal behaves is very much affected by what its opponent does. Animals assess putative opponents, spy on others, modify their behaviors when they are watched, form alliances to subdue rivals, and more. Studies of the dynamics of power show just how complex their strategies can get.

For three weeks in March in 1990, two University of Hamburg ethologists, Dierk Franck and Alexander Ribowski, sitting on the shores of various creeks and streams in Veracruz, Mexico, gathered data on aggression in a group of nearly a hundred fish known as green swordtails. From the attacks and retreats, the nipping and flashing, the lateral displays and the body rams they observed, Franck and Ribowski discovered that these fish form tidy dominance hierarchies, but scientists weren't sure how they did it.

A few years later, when Ryan Earley joined my lab as a doctoral student, he set his sights on probing even deeper into the nature of power in swordtails. After hundreds of hours of watching males in the lab, he felt certain the swordtails were conducting reconnaissance or what in the animal behavior literature is called eavesdropping. Eavesdroppers use information they glean from watching other fights and change their assessment of the fighting abilities of those they have gathered intel on.

In an ingenious experiment involving one-way mirrors, Earley discovered that swordtail spies avoid interacting with the winner of a contest they have watched. When it comes to interacting with males they've seen lose, the fish follow an intriguing rule: if a loser puts up relatively little resistance, go after him; but if he has moxie and puts up the good fight before eventually capitulating, stay the hell away from him. The swordtail's intelligence gathering and the way it uses that information shows nicely that natural selection sculpts subtle and complex strategies used during power struggles, even in an animal whose brain could sit comfortably on the head of a pin.

Other animals strategically change how they behave depending on who's watching them, trying to shift the balance of power their way. An intriguing example of this strategy comes from the ravens that Thomas Bugnyar, Georgine Szipl and their colleagues have studied at the Konrad Lorenz Field Station near the village of Grünau in the Austrian Alps.

From the perspective of the ravens at the station, a human audience to power struggles is not worth paying mind to, but an audience made up of other ravens definitely is. Victims of aggression often give off defensive calls that entice raven audience members to come to their aid. But Bugyner and Szipl sensed there was an added layer of complexity at work as well. They videotaped victims giving a defensive call, and then when watching the tapes, noted not just the duration and number of calls but the identity of other ravens within 25 meters of the victim. It turns out that ravens on the wrong side of a fight temper their defense calls depending on who's watching and listening. Victim call rates were highest when potential allies—either relatives or long-term associates (friends)—were in the audience. Even more remarkably, victims reduced their call rate when their opponent had potential allies in the audience: no sense drawing even more attention to an unfortunate predicament when it might make matters even worse.

For the dwarf mongooses at Queen Elizabeth National Park in Uganda, the most intense power struggles occur between groups. Michael Cant and his colleagues wanted to know why, and found that it all starts because genetic relatedness in mongoose groups builds up over generations. This arrangement can lead to inbreeding, but female dwarf mongooses have found a simple yet clever way around this problem: look for mates in nearby groups. When a female leaves in search of a mate, males from her group follow her, which often leads to an all-out battle between males from the two groups. These are not pretty affairs; they often involve many casualties, including deaths, among males. But the female who came looking for a mate will often find one while the males from her group are otherwise engaged.

Spying swordtails, wily ravens and scheming mongooses are just three examples of the means by which animals make their power plays. Power struggles take place on the land, underground, in the air and in the water on every continent and have been studied in detail in hundreds of species, including hyenas, caribou, chimpanzees, bonobos, dolphins, deer, horses, field mice, ravens, skylarks, white-fronted bee-eaters, copperhead snakes, wasps, ants and cuttlefish.

The more we learn, the more we discover the myriad ways that animals continue their unceasing pursuit of power.

About the Author

Lee Alan Dugatkin is a professor of biology at the University of Louisville. His newest book is Power in the Wild: The Subtle and Not-So-Subtle Ways Animals Strive for Control over Others *(University of Chicago Press, 2022). Twitter: @ Leedugatkin; Facebook: facebook.com/lee.dugatkin/.*

The Surprising Architecture in Bees' Honeycombs

By Saugat Bolakhe

Charles Darwin described bees' ability to build perfect honeycombs as "the most wonderful of all known instincts." Each hexagonal cell is so precisely constructed and so neatly arrayed that a comb is a visual treat. Now new research into how honeybees incorporate different cell sizes into a flawless-looking grid–and seamlessly merge combs built from multiple directions at once–reveals remarkable adaptability.

Adjacent honey storage cells are typically uniform in size, but bees must build some larger cells for rearing drones and smaller ones for workers. They also have to align and join comb grids being constructed from separate starting points. Understanding how bees accomplish all of this so elegantly "is an old quest," says Auburn University behavioral ecologist Michael Smith.

For a study in the *Proceedings of the National Academy of Sciences USA*, Smith and his colleagues measured 19,000 comb cells from 12 colonies of Italian honeybees. Using automated image analysis, they examined the centers and vertices of these cells to find variations in their shapes and sizes. "We just couldn't extract the precise measurement of thousands of cells until we had this new automated method," Smith says.

The researchers found that bees made clever adjustments to transition between sizes and merge combs as cleanly as possible. The insects utilized patterns of irregular shapes (mostly pairs of heptagons and pentagons) and manipulated comb cells' size and orientation so craftily that their ability can be considered "a true architectural skill," the researchers write.

"The hexagonal grid structure of a honeycomb–constructed by a leaderless collective of hundreds of bees–lends itself to speculation that robotic, innate behavior must be at work," says Queen Mary

University entomologist Lars Chittka, who was not involved with the study. “But a simple robot does not have such a level of adaptability and rate of error recovery.”

Raghavendra Gadagkar, who studies insect behavior at the Indian Institute of Science, Bangalore, and who was also not part of the study, suggests that examining how bees coordinate comb building can help advance robotics. “Imagine,” he says, “if we could program such wits of honeybees in real-world robots.”

Play Is Serious Business for Elephants

By Caitlin O'Connell

It was late afternoon in the winter scrub desert within Namibia's Etosha National Park when I spotted a family of elephants on the southern edge of the clearing. I was scanning the horizon from the observation tower where my colleagues and I conduct our research at Mushara water hole. Wind had deterred elephant families from visiting the water hole earlier–it interferes with their efforts to keep tabs on one another vocally–but with the air now still, our first customers of the day had finally appeared.

Judging from how many trunks were stretched high, sampling the air, the group was itching to break cover and run for the water. The young males were particularly anxious to get going. Not only were they thirsty, but they had a lot of sparring to catch up on. As winter wears on, the environment dries out, and elephants have to venture farther from water to find enough to eat. Several days may pass before they can return to the water hole for a drink and a reunion.

I could see why this group was holding back, however. Another elephant family was amassing in the southeastern forest and heading our way, and the adult females were wary. They stood with their feet firmly planted, ears held straight out, as they sniffed what little remained of the prevailing wind for any potential danger. Not only would exiting the security of the forest expose the family to predators, but an encounter with a higher-ranking elephant family could result in an aggressive interaction. For the youngsters in the group, however, more families meant more opportunities to play. So after thoroughly assessing the clearing, the matriarch gave the word with a rumble and an ear flap, and the family began its approach to the water.

Late afternoon is my favorite time of day during our field season in the austral winter–the air cools fast as the sun sinks low in the sky, painting the elephants a radiant pink. My colleagues and I stand in the observation tower with a celebratory drink in hand, our binoculars trained on the horizon, hoping for a sunset visit like this one from one of our beloved resident families. During these daily visits, I always learn a new lesson about elephants–particularly when they play.

I have witnessed the important role of play in calf development and family politics by watching members of my favorite elephant groups frolic at this water hole at sunset. These often chaotic observations inspired me to want to understand more about how animals play and what advantages this behavior might confer, not just to elephants but to all social creatures, including humans. It turns out that play, like other forms of interaction, has rules of engagement. And it is essential for developing the physical and cognitive faculties that animals need to survive and reproduce.

Rules and Regulations

People tend to think of play as an activity one engages in at one's leisure, outside of learning important skills needed to succeed later in life, such as hunting, mating, and evading predators. But although playing is fun for all involved–and fun for those who are watching–play behaviors evolved as ritualized forms of survival skills needed later in life, providing the opportunity to perfect those skills.

Engaging in play allows animals to experiment with new behaviors in a protected environment without dangerous consequences. The unwritten code of conduct surrounding play lets them explore many possible outcomes.

Animals learn the rules of engagement for play at a very young age. Among dogs, the bow is a universal invitation to engage in silliness that triggers the same bowing down and splaying of the front legs in the receiver of the signal–inevitably followed by chasing and pretend biting. Chimpanzees and gorillas motivate others to romp by showing their upper and lower teeth in what primatologists refer to as a play face, which is comparable to human laughter.

When a young male elephant wants to play with another male of similar age, he holds his trunk up and presents it to the other as an invitation. Most often his next move would be to place his trunk over the other's head, which in adults signals dominance but in calves is guaranteed to precipitate a spirited sparring match. These encounters run the gamut from gentle shoving to intense headbutting and pushing back and forth with trunks entwining and tusks clacking. The fun continues for seconds to minutes for youngsters; for older teens and young adults, it can go on much longer. The sparring matches provide bulls with the opportunity to test their fighting ability so that they might successfully compete for a female when they reach sexual maturity and enter the hormonal state of musth around the age of 25.

When a young male elephant is feeling particularly adventurous, he may venture far away from Mom's protection to invite a distant relative to spar. If his foray takes him too far away or if a spar turns unexpectedly rough, the brave calf will lose his nerve and often will run quickly back to Mom's side with ears flapping and trunk yo-yoing as he retreats.

Occasionally an older sister will oversee a play bout between youngsters. These ever watchful siblings form part of an extended caretaking network that facilitates play, but its members also will intervene if a calf crosses an invisible bloodline and gets deflected with a trunk slap by an overly protective, high-ranking mother.

Forms of Play

Scholars of animal behavior recognize three main categories of play. The first is social play, which is any kind of antic that involves others. The second is locomotive play–including running, walking, jumping and pouncing–which facilitates lifelong motor skills. In prey species, locomotive play helps perfect predator-avoidance tactics such as the springbok's "pronking" high into the air while running as a herd and landing in unpredictable spots. In elephants, it hones predator-avoidance skills, as well as strategies for escaping an aggressive suitor or a competitor looking to inflict a mortal wound. Conversely, young predators such as lion cubs use locomotive play to sharpen their hunting ability. Chasing and tripping littermates and then giving them a good chew on the spine or throat are rehearsals of the skills needed to catch prey animals and dispatch them by severing their spinal cord or choking them.

Many species, including our own, engage in the mock-fighting variety of locomotive play, which allows them to test their strength in a safe environment where everyone understands the rules. A playful spar in elephants is just like an arm wrestle between human peers. When play becomes more elaborate and determined, it turns from an arm wrestle into something akin to martial arts, allowing both participants to practice skills and develop innovative solutions that could help them avoid mortal combat later in life. Play fighting also provides opportunities to test boundaries, gauge who can be trusted and learn important body language.

The third main category of play is object play, which incorporates objects from the environment into the cavorting. For an elephant, this object might take the form of a stick or branch that the elephant explores, carries or throws with its trunk. In captivity, elephants enjoy playing with balls or hauling inner tubes around for fun. Alternatively, the object could be another animal, such as a zebra or giraffe, that offers an irresistible opportunity for a chase. In one case, a four-year-old male calf named Leo taught his baby brother, Liam, just how fun such a chase can be, leaving Liam scrambling to keep up with Leo's charge as a giraffe made a quick escape.

Two other forms of play have only been documented in great apes, including humans. One of these, game playing, combines social, locomotive and object play. Sports such as soccer, field hockey, lacrosse and polo are examples of traditional games that became formalized as sports with specific sets of rules (among nonhuman great apes, only captive individuals raised in human contexts play formal games). The other variety of play that appears to be unique to great apes is make-believe. For example, a wild chimpanzee may carry around a small log, pretending it is an infant. A human child might play with an invisible toy or set up an invisible barrier that they want adults to acknowledge.

Not Just Fun and Games

Play provides an environment for experimenting with risk. When a lion cub deliberately gives up some control over its body, it puts itself at a disadvantage, allowing others to succeed in pouncing on it. Marc Bekoff of the University of Colorado Boulder and his colleagues have proposed that play increases the versatility of movements used to recover from a loss of balance and enhances the ability of the player to cope with unexpected stressful situations. The goal is not to win but to improve skills, sometimes by self-handicapping.

Once a cub has been tackled by its littermates, roles might reverse such that a littermate handicaps itself, allowing the other cub to tackle it in return. Self-handicapping is risky and requires trust, but it is a great way to develop strength and agility. It is also an important exercise in building cooperation. In the Sawtooth wolf pack raised by Jim and Jamie Dutcher in the Sawtooth Mountains in Idaho, the dominant wolf would slow down to allow a close companion that happened to be a subordinate to catch up and tackle him. In elephants, on a number of occasions I have seen older male calves crouch down to allow a much younger calf to spar with them. This is akin to an older brother handicapping himself during an arm wrestle by not using all of his strength to let his little brother win.

Being silly is another important aspect of play, one that gets us outside our comfort zone and forces us to test new strategies. Silliness in our movements, behavior and even language helps us think much more broadly and creatively. Problem-solving derived from the silliness of play has been demonstrated in many species and even in robots. When computer scientist Hod Lipson of Columbia University gave his artificial-intelligence robots a chance to play–by dancing around in random movements–they outperformed other robots when challenged with the unexpected. The positioning information garnered from moving around randomly led one robot to come up with creative solutions for maintaining its balance after losing a limb. Likewise, when sea lions play in the surf, they often project themselves high into the air midway down the face of monster waves, like those that roll into Santa Cruz. These are just the kinds of behaviors needed to avoid an attack by a great white shark—their primary predator apart from killer whales and humans.

Play also builds trust. Thomas Bugnyar of the University of Vienna in Austria and his colleagues found that ravens pretend to cache highly valued food items and then watch how other ravens respond, apparently to determine whom they can trust. Learning how to differentiate competitors from likely reliable collaborators early on has obvious advantages, whether one wants to gain allies or build a coalition within a group—or repair broken relationships.

Families Reunited

"Incoming from the southeast!" I called out from the Mushara tower as my elephant field team narrowed in on what looked like a dusty line of pinkish-gray boulders amassing on the edge of the clearing one afternoon during our 2018 field season. The search for identifying features began. A missing tusk, a notch in the bottom of the left ear, or a V-shaped cut in the top of the right ear would give the family away. Whoever identified the elephant family first would get an extra sundown drink.

That day the incoming family turned out to be the Actors. It was our first sighting of the group that season, and we were excited to see a new addition to the family: high-ranking Susan, identified by her daggerlike left tusk, had a new male calf, Liam. And low-ranking Wynona, who was missing her left tusk, had her two-year-old calf Lucy in tow. We had been following the contentious dynamic between these two mothers very closely over the years, particularly during the 2012 season when each had a calf—Leo and Liza, respectively.

Susan had relentlessly tormented Wynona all the way up to the end of her pregnancy, aggressively charging her whenever she got close to the water to drink. The tension was so high that when

Wynona broke away from the family to give birth, surrounded by her daughter Erin and their calves, I worried for her baby's life if a reunion were to take place. Sure enough, there was no fanfare and no reunion that we witnessed to present her new baby to the rest of the family. I assumed then that Wynona's days as a member of the Actor family were numbered.

As predicted, Wynona did separate from the larger family and became the matriarch of her own core family. It went on like that for four years until the arrival of Wynona's newest baby, Lucy, in 2016 yet again changed the dynamic of the larger extended family group. Play appeared to be an important contributing factor in reuniting the family.

Lucy's older sister, Liza, had been a shy baby who stuck to her mom and her very close relatives. Wynona timed her movements to avoid too much overlap with the larger family group when they went to Mushara water hole to drink. They tended to be one day behind or ahead of the Actor family, usually behind. On the rare occasion that they did overlap just at the end of the extended family visit, Liza did not stray to interact with the larger family. And who would blame her? Susan was right there with a quick jab with her dagger tusk or a trunk slap, whichever was more convenient, making it clear that the low-ranking babies had no place on the playground with royalty. There was hardly a chance for calves of Wynona's small but growing family to get to know members of the extended family.

Lucy changed all that. From the start, she was quite the extrovert. Maybe being born into a very small family made her all the more curious and excited by the opportunity to engage with the extended family on the infrequent occasion of their overlapping. And she was not deterred by the admonishments of high-ranking moms within the extended family, much to the seeming annoyance of the ever watchful Susan.

Now the two-year-old Lucy knew just how to run through adults' legs and out of trunk's reach, navigating potential minefields and dodging her mom's attempts to rein her in. She behaved more like Susan's calf, Leo, who was her older sister Liza's contemporary.

When we scored Leo's distance from his mom at the water hole, he always had a much higher score than Liza. We had assumed that was attributable mainly to his sex and the male elephant's early experiments with independence. But the arrival of Lucy showed us that the story was not that simple.

Lucy spent a lot of time a great distance away from her mom and played with calves of mothers of all ranks. When it came time to leave the water hole and go in separate directions, as dictated by the prevailing family politics, Lucy made that impossible. She was so busy playing with other calves that there was no extracting her, leaving Wynona no choice but to modify her behavior.

Instead of continuing on her premeditated departure route, in the opposite direction from the Actor family, Wynona, her eldest daughter Erin and their calves turned around and followed the rest of the family so that Wynona did not risk losing her new calf. There was no guarantee that the other mothers would protect Lucy, much less allow her to suckle, as that would mean fewer precious nutrients for their own calves. But by 2018 Wynona was fully reintegrated into the Actor family, whether she wanted to be or not.

Every time I see this dynamic unfold, it makes me smile. How often is it the case in our own families where grudges of older generations are put aside because of the bonds forged by the next generation through play?

Play should be on our daily agenda. Smiling and laughing are contagious behaviors that facilitate bonding, are curative and, most important, do not have to take up much time. The next time you feel like you are too busy to play a frivolous game at work or you don't want to face that family reunion, make the time and muster the will. You might be surprised at the outcome, whether it be a better idea for a pitch meeting or the dissolution of a long-standing barrier between you and a contentious relative thanks to a good giggle.

Our highly adaptable and innovative nature is rooted in play. I am grateful to my favorite elephant, Wynona, and her daughter Lucy for reminding me that there is always something new we can learn from it—and that we are never too old to internalize those lessons.

A good romp can pay off in ways I hadn't anticipated. It forges new bonds, reunites divided families, improves coping skills and overall health, and facilitates cooperation and innovation. Given all these benefits, how could we afford not to play?

About the Author

Caitlin O'Connell is a behavioral ecologist at Harvard Medical School. Her research focuses on elephant behavior and communication. O'Connell's latest book is Wild Rituals: 10 Lessons Animals Can Teach Us about Connection, Community, and Ourselves *(Chronicle Prism, 2021).*

How This Zombie Fungus Turns Cicadas into Horror-Movie Sex Bots

By Rachel Nuwer

Matt Kasson, a forest pathologist and mycologist at West Virginia University, has spent the past week chasing cicadas across northern Virginia. Each night he has watched pale nymphs emerge from the earth like ghostly apparitions and then make their way into trees, where they will soon molt to become a winged adults.

Both the cicadas and Kasson have been waiting for this moment–but not for the same reason. After 17 years underground, billions of Brood X periodical cicadas, as the insects are known, are emerging into the light across the eastern U.S. this year to find mates and launch the next generation. Kasson, on the other hand, has come for an equally awe-inspiring yet very different facet of one of nature's most epic life cycles: "the big fungal show," as he puts it. It is one wild performance in which a parasitic fungus turns some of the insects into singled-minded sex machines–after causing their genitals to fall off. The phenomenon is well documented, but scientists such as Kasson are working to figure out *how* the freaky fungus hijacks its host.

An estimated 3,000 species of cicadas live around the world, and a diversity of parasitic fungi have evolved alongside them. Kasson studies one exceptionally peculiar species called *Massospora cicadina*. It only infects so-called periodical cicadas such as Brood X–those that emerge every 13 or 17 years.

M. cicadina's spores have also patiently spent the past 17 years in the soil, awaiting the Brood X nymphs. The immature insects become infected when they come into contact with the spores on their way to the surface, and the fungus grows out of sight for the first seven to 10 days as its host molts into an adult.

After that deceptive period of quiet, however, the infection begins to manifest–with horrifying results. Males' abdominal plates begin to slough off, revealing a chalky white plug made of fungal spores. The spores can come off when an insect walks or flies. But mostly, they spread through sex attempts.

Healthy male cicadas sing to attract females, which indicate their receptiveness to mating by flicking their wings. A female unfortunate enough to fall for an infected male's siren song will not only fail to get her eggs fertilized (the male's genitals have long since fallen off) but will also become infected herself, essentially making *M. cicadina* a sexually transmitted disease.

An infected female's abdomen tends to remain more or less intact because they have stronger muscular tissue in place for laying eggs. But that does not mean females have it easy. If a male mates with an infected female, her fungus-ridden genitalia usually break off–and get carried away by the male. "Basically, the male has this permanent ornament attached to him with a legacy of deeds past done," Kasson says. Some infected males also flick their wings like a female, attracting other hapless males that become infected when they attempt to mate.

Much of Kasson's research entails figuring out how in the world the fungus compels cicadas to become sexual zombies for spreading its spores. In 2019 he and his colleagues reported in *Fungal Ecology* that one compound in particular, cathinone, seems to be key for turning *M. cicadina*–infected hosts into sexual automatons. Now, in Virginia, he is screening cicadas for fungal infections, looking for sample insects to study in the lab. "Also," Kasson adds, "I have to admit I wasn't going to miss the opportunity to just spend time with Brood X."

Cathinone is a well-known, naturally occurring stimulant in the amphetamine family. Scientists had previously pinpointed it as an active ingredient in khat, a plant native to the Horn of Africa and the Arabian Peninsula, where people sometimes chew it for a buzz. Synthetic cathinone is also a key component in pharmaceuticals prescribed for attention deficit hyperactivity disorder and in illegal drugs called "bath salts." "We aren't seeing face-eating cicadas," Kasson says, referring to a notorious 2012 Florida incident that media reports blamed on bath salts. "But we can learn something about how the fungus might be working from what these medicines are used for."

Medications that contain synthetic cathinone, for example, can suppress appetite and increase the ability to focus. So it is possible the fungus alters host behavior to its advantage by making cicadas single-mindedly pursue mating rather than feeding or resting–thus spreading the spores to more mates.

Interestingly, Kasson and other researchers have also found evidence of psilocybin–the psychedelic ingredient in magic mushrooms–in another *Massospora* species that only infects annual cicadas in the western U.S. "Psilocybin and amphetamines are obviously two very different types of drugs, but we think these alkaloids could be having a bit of redundancy in the ability to suppress appetite, increase stamina and prolong wakefulness," Kasson says.

He suspects that the presence of these compounds in *Massospora* species represents an example of convergent evolution, in which a feature independently evolves in different species to perform a similar function. Cathinone's occurrence in khat plants, however, does raise one intriguing question: Could these plants contain hidden fungal partners that are actually the engineers behind the compounds? As Kasson points out, this would not be the first time that credit for a plant's apparent psychoactive properties turned out to belong to a fungus: in 2004 scientists discovered that an LSD-like compound found in morning glory was not produced by the plant itself but by a fungus that lives inside it.

Kasson hopes to further investigate this and other questions in future studies. In particular, he would like to dose healthy cicadas directly with cathinone to see how it alters their behavior. He is also interested in looking at the dozen or so other *Massospora* species out there to see what interesting compounds they may hold–something that could even be useful for pharmaceutical drug discovery.

Speaking of drugs, however, Kasson emphasizes that people looking to get high should not turn to infected cicadas–something that he has been concerned to see discussed in online forums. It would take hundreds or even thousands of cicadas to produce a dose high enough to have an effect on humans, he says. And cicadas also contain more than 1,000 other compounds, some of them known toxins. "You'd have to eat an awful lot of cicadas to have some effect from the amphetamine that's in there," agrees Chris Simon, a phylogeneticist at the University of Connecticut. "You'd be sick long before you ate enough."

That said, both Simon and Kasson say people should not worry much about pets or kids coming into contact with infected cicadas. For starters, it may be difficult to find them. Only about 5 percent of cicadas tend to be infected in the first place. And although Brood X consists of billions of insects, they are dispersed over a large geographical area. "I wouldn't lick one myself," Kasson says. "But if your animal eats one or two infected cicadas, that's not going to be a problem."

About the Author

Rachel Nuwer is a freelance science journalist and author who regularly contributes to Scientific American, *the* New York Times *and* National Geographic, *among other publications. Follow Nuwer on Twitter @RachelNuwer.*

Section 2: Amazing Animal Anatomy

Octopuses Redesign Their Own Brain When They Get Chilly

By Rachel Nuwer

Octopuses are among the smartest animals on the planet–and some of the strangest. They have about the same number of neurons as dogs, but more than half of those cells are distributed across the slippery cephalopods' eight arms rather than contained in a central brain.

As researchers report on June 8 in *Cell*, the neural anomalies only get more bizarre from there. Octopuses, they found, have the ability to recode their neurons in response to temperature shifts so those cells produce different proteins. Like people adjusting their clothing to match the weather outside, octopuses edit their RNA, which is a genetic molecule that carries DNA's instructions to produce proteins–the workhorses of cells. The researchers suspect those "brain edits" help octopuses adapt to heat or cold when the seasons change. And they do this to "an extraordinary extent," says co-senior author Joshua Rosenthal, a biologist at the Marine Biological Laboratory in Woods Hole, Mass.

RNA editing happens when an outside force activates certain enzymes inside the body's cells that then make chemical changes to the RNA. Depending on the changes, the cells produce different forms, or isoforms, of proteins. Because RNA is a transient molecule, any changes made to the genetic information it carries will not be permanent–a feature that, in theory, makes it a powerful tool for on-the-go acclimation to changing environmental conditions.

In humans, RNA editing affects protein production in fewer than 3 percent of genes, whereas Rosenthal and his colleagues previously found that all sophisticated species of cephalopods–that is, all cephalopods other than nautiluses–can recode the majority of neural proteins.

The researchers wanted to follow up on that earlier work to see what factors might drive RNA editing in cephalopods. They started with temperature change because it's a straightforward environmental condition that fluctuates seasonally, or even daily.

The scientists first collected a dozen wild California two-spot octopuses (*Octopus bimaculoides*), a species whose genome has already been sequenced. They acclimated the animals to tanks that held either warm or cold water. Several weeks later, the researchers examined around 60,000 previously identified sites in the animals' genomes where enzymes edit RNA. They found that about one third of those sites had changed and that these changes happened quickly, on the scale of hours to a few days. "We expected to see here and there a few sites that had changed, but no, this was something very global," says co-senior author Eli Eisenberg, a physicist at Tel Aviv University, whose research focuses on RNA.

Almost all the changes were cold-induced, the team found. And among those changes were ones that encoded for specific classes of proteins involved in cell membranes, functions of synapses (which transmit neuronal signals), autophagy (programmed cell death) and the binding of calcium (which plays various roles within neurons). The researchers confirmed that the isoforms created via the edited RNA had altered functions, but "we do not know yet how these thousands of changes, or some of them, promote adaptation," Eisenberg says. "Understanding the global effect of all the changes in concert is left for future studies."

Finally, the team also collected wild octopuses–including another closely related species, Verrill's two-spot octopus (*Octopus bimaculatus*)–in summer and winter. Those individuals, the researchers found, exhibited the same temperature-tracking RNA-related changes as the California two-spot octopuses they had tested in the lab.

Unlike humans and other mammals, octopuses can't regulate their own temperature. So the researchers suspect that RNA editing plays a role in protecting the invertebrates' neurons from temperature fluctuations. "The organism chooses to express different isoforms, and each one is better in its own condition," Eisenberg says. "There's not even a single example of that happening in mammals."

Heather Hundley, a biologist at Indiana University Bloomington, who was not involved in the research, calls the new paper "a true tour de force."

"This work adds to the growing body of data demonstrating that RNA editing can be dynamically regulated," she says. "While previous work has primarily focused on regulation during development and disease, this work demonstrates that RNA editing can serve as a molecular means to acclimate protein function in response to environmental changes in temperature."

Rosenthal, Eisenberg and their colleagues plan to continue this line of study by investigating additional factors that might trigger changes in protein production, such as pH and oxygen levels or the social environment. "How broad is this ability to acclimate through RNA editing?" Rosenthal says. "There's a lot of interesting work to be done."

About the Author

Rachel Nuwer is a freelance science journalist and author who regularly contributes to Scientific American, *the* New York Times *and* National Geographic, *among other publications. Follow Nuwer on Twitter @RachelNuwer.*

How Do Birds Know When to Migrate?

By Meghan Bartels

For a migratory bird, good timing is a matter of life or death. Every spring countless warblers, geese, hummingbirds, and more—one in five bird species, all told—leave their winter hideaway on a long trek that eventually leads to mating, nesting and hatching chicks. Knowing when to leave is a delicate calculation for these animals, however. They need to reach breeding sites in plenty of time to successfully reproduce. But if they arrive too soon, winter may not have quite loosened its grip, which puts the birds at risk of starvation.

To time this arrival for the highest likelihood of conditions favorable to reproducing, a migratory bird tracks each day's light length, or photoperiod, says Marilyn Ramenofsky, a behavioral endocrinologist at the University of California, Davis, who specializes in birds. That tracking happens instinctively in the animals, which use special photoreceptors in their brain to detect light. These receptors are similar to those that govern humans' circadian rhythm. Though scientists don't fully understand these light-sensing receptors, one team identified certain brain cells in quails that could directly detect light, according to a study published in 2014 in *Current Biology*.

In general, once those photoreceptors register that the days are lengthening in late winter and early spring, they trigger activity in a host of body systems.

Birds molt, for instance, shedding their worn feathers and growing new ones. Their hormones change, influencing their digestive patterns and leading them to eat more, put on weight and build up flight muscle. In some birds, once they've eaten enough, a digestive hormone called ghrelin begins to make them restless and nudges them to take flight. And the birds' reproductive tracts begin to change in order to ensure that they will be fertile by the time they breed.

"It's really a cabal of different endocrine systems that are shifting the bird from winter survival to migratory preparation," Ramenofsky says.

When all the pieces fall into place, birds are well prepared to raise their chicks while making the most of the long days and the seasonal boom in bugs, berries and other resources their summer homes offer.

Each bird undergoes these seasonal changes a little bit differently. "They're all sort of like little machines in terms of getting turned on, but there are individual decisions," Ramenofsky says.

For example, a higher-quality winter habitat can delay an individual bird's departure, says Ana M. González, an ornithologist at the government agency Environment and Climate Change Canada.

Weather conditions can be particularly important for timing, González says. She notes that birds may wait to depart until a storm passes. In addition, birds that are already en route can speed up or slow down their journey in response to the weather, which can help them time their arrival at breeding sites, González says.

Understanding migration is also complicated because the phenomenon comes in many forms. Some birds, such as Savannah Sparrows and American Woodcocks, migrate only short distances. Others, such as Cerulean Warblers and Red Knots, begin their prebreeding migratory journeys in the Southern Hemisphere, where days are getting shorter, not longer, and then cross the equator to reach North America.

Both scientists emphasized that researchers don't yet understand the nuances that shape each species' migration, much less each individual bird's migratory timeline. "There's a lot of variation [from] species to species and even within a species," González says.

The return journey after breeding is even more of a puzzle. A day's photoperiod seems to tie to the initiation of breeding steps, but by the time of the year when birds are returning, they are no longer sensitive to changes in day length, she adds. "The actual cue, we don't know," Ramenofsky says.

Fortunately, scientists' tools for understanding migration are improving. Technology for tracking birds across long distances, particularly via satellite, is getting better. And community science initiatives–particularly the eBird program, which gathers reports of what species birders see where and helps scientists trace migratory pathways and schedules–are giving researchers many more eyes on birds.

About the Author

Meghan Bartels is a science journalist and news reporter for Scientific American *who is based in New York City.*

How Bears Hibernate without Getting Blood Clots

By Jack Tamisiea

Despite spending more than half the year hibernating in frigid temperatures, brown bears (*Ursus arctos*) remain remarkably healthy. These idle giants rarely suffer from maladies such as blood clots, which can occur in humans experiencing only temporary bouts of immobility and can be deadly. "Patients who come into the hospital with a broken leg are usually prone to develop thrombosis," says Tobias Petzold, a cardiologist at the German Center for Cardiovascular Research. "But in contrast, brown bears, which are lying around for a couple of months, do not develop clots."

In a new study published on Thursday in the journal *Science*, Petzold and his colleagues pinpointed a protein that helps the hibernating bears avoid dangerous clotting that can prevent blood flow during their long winter slumber.

To determine how dormant bears keep blood pumping during months of torpor, Petzold and his colleagues partnered with biologists studying a population of brown bears in Sweden. The biologists collected blood samples from 13 hibernating bears in their dens during the winter. In the summer the biologists collected additional blood from the same bears after shooting tranquilizer darts at them from a helicopter.

The researchers then ran these samples through a battery of blood tests. They discovered that the quantities of more than 150 proteins varied greatly between the blood of hibernating and active bears and homed in on the proteins in platelets–the components of blood that cause clotting. The platelet protein with the largest disparity between active and hibernating bears was heat shock protein 47 (HSP47).

HSP47 recruits an enzyme called thrombin that helps platelets stick together and form clots. In active bears, HSP47 helps patch up cuts and stop bleeding. But hibernating bears, which are safely snug in their dens, have little use for this clotting protein. On average, hibernating bears' platelets produced 55 times fewer HSP47 proteins than those of active bears.

"We hadn't heard much about this protein, and we were completely surprised to discover that it has such a large impact," says Manuela Thienel, the paper's lead author, also at the German Center for Cardiovascular Research. Thienel says that decreasing the levels of this particular protein likely reduces the platelets' penchant for clumping together and restricting blood flow.

Mirta Schattner, a biologist at the Laboratory of Experimental Thrombosis and Immunobiology of Inflammation at the Institute of Experimental Medicine in Argentina, agrees. "This cellular interaction [of HSP47 proteins] is beneficial when limited and is [also] very important for containing infections" and inflammation, says Schattner, who was not involved with the research team but wrote a commentary on the new study in *Science*.

To determine if a similar mechanism prevents blood clotting in chronically immobile humans, the researchers compared blood samples from patients who had suffered debilitating spinal injuries with samples from their active counterparts. Like that of hibernating bears, the blood of chronically immobile patients had fewer circulating HSP47 proteins. "Downregulation of HSP47 during hibernation of bears or chronically immobilized patients reduces thromboinflammation" and thus reduces the risk of blood clots, Schattner says.

The team also collected blood from several healthy individuals before they participated in a month-long spaceflight simulation study run by NASA and the German Aerospace Center. After the participants experienced 27 days of head-down bed rest, the researchers collected another round of samples to see how the prolonged immobilization had impacted their physiology. The team discovered that the participants' platelets began to produce significantly less HSP47 over the course of their experimental immobilization.

The researchers found a similar pattern in the blood of mobility-limited pigs and lab-bred mice. They posit that reducing the expression of HSP47 proteins may be a mechanism used across mammals to prevent clotting during prolonged periods of rest. But this approach appears to take time to develop–in humans, people experiencing shorter-term immobility from illness or injury are more susceptible to clotting than chronically immobile people with spinal cord injuries.

Petzold says the new findings highlight how understanding the biology of bears and other animals can provide insights into how the human body functions. And they could potentially inspire fine-tuned treatments for preventing clotting in patients experiencing temporary immobility. "By looking at how hibernating bears in nature prevent clots, we now know that humans do the same thing to prevent blood clots," he says.

About the Author

Jack Tamisiea is a science journalist based in Washington, D.C., who covers natural history and the environment. Follow Tamisiea on Twitter @jack_tamisiea.

Giraffes vs. Blue Whales vs. Dinosaurs: Which One Builds Its Nervous System Fastest

By Douglas H. Smith, Jeffrey M. Rodgers, Jean-Pierre Dollé, Mathew J. Wedel

Over the course of evolution, immense megafauna have roamed the lands or swum in the seas. The growth of these creatures early in their life is typically quite rapid. It has to be. They need to grow fast or be eaten. Extensive studies on megafauna have addressed the unique challenges of supporting and moving such massive bodies. But the greatest and largely ignored obstacle to extreme growth for both terrestrial and aquatic megafauna may involve the rapid development of their nervous system.

Unlike all other cell types, neurons do not increase tissue volume by cell division but rather by expanding the volume of the cells themselves. Early in development, neurons sprout nerve fibers, or axons, that extend from their cell body—which houses the nucleus and other structures—by using chemical and physical cues to navigate and slowly grow toward target cells, typically no more than a few millimeters away. Once the target is reached, this well-studied first phase of axon growth culminates with formation of a junction with another neuron, known as a synapse.

But the growth and elongation of axons does not stop there. It continues by relying solely on mechanical forces that can extend the fibers at seemingly impossible rates. This poorly studied second phase, called "stretch growth," continues as the distance between the neuronal cell bodies and their targets increases.

As a creature grows, axons engage in a tug-of-war contest. Tension from pulling on the rope-like axons threatens to rupture the fibers because they only stretch so far. Tautness is relieved by stretch growth that continuously adds length to axons. Mechanical stretching of the axons as a body grows is ubiquitous throughout nature and thought to drive the formation and organization of axon tracts that make up the white matter in the brain and spinal cord and peripheral nerves outside the brain. The growth of vertebrae during development stimulates the growth of axons in the spinal cord. Axons in peripheral nerves extend alongside the growing long bones and other rigid structures. Still, many mysteries remain about this process, and just how fast and far axons stretch in both terrestrial and aquatic megafauna has yet to be settled.

A Few Centimeters a Day

One obvious example of massive growth is the blue whale, which can reach colossal lengths of up to 30 meters at maturity. Using recorded measurements of the whale's body length at different times during development, we previously calculated the peak stretch growth rate of axons that span the spinal cord as an astonishing 3.4 centimeters per day. This rate stands in marked contrast with those described in neuroscience textbooks. Indeed, a crucial axon building block called neurofilament protein moves down these nerve fibers at only up to a few millimeters each day. This approximately matches the extension rates of axons sprouting in culture and regenerating from injury, both of which add to cell volume at the ends of extending axons.

Even at this speed, it would still take years for proteins to complete the journey down the longest axons in maturing whales. Recent computational modeling analyses, however, suggest that the new proteins for axon stretch growth are extruded, or pushed out, from the neuron cell body so that they don't have to make the increasingly longer trip to the end of the cell to extend axon length.

Even here, the pace of axon building in the blue whale seems impossible to maintain. Indeed, every 3-cm increase in axon length is calculated to add more than double the volume of the neuronal cell body to the axon each day.

Remarkably, this increase in size rivals the peak growth rate of aggressive cancer cells that double in number each day. In the blue whale, this process reaches astonishing proportions–spinal cord axons can extend 24 meters in length from brain stem to tail. Although axons are quite thin in diameter, such a long extension ultimately results in their volume being more than 1,000-fold greater than that of their cell bodies. This bizarre geometry of neurons with very long axons potentially represents one of the largest cells by volume for any mammal. For routine cellular maintenance, it is hard to imagine how the neuronal cell body can provide sufficient proteins and other cellular supplies for such a large axon volume that extends so far away from the cell body. It has been suggested that protein synthesis might occur along axons themselves to overcome the challenges in production and distribution, but we still have much to learn about this unusual cellular process.

Enter the Dinosaurs

Despite the many mysteries of its mechanisms, extreme axon stretch growth obviously works well for the immense blue whale. Yet while 3.4-cm-a-day axon growth is impressive, the blue whale has steep competition from other megafauna. Enter a dinosaur known as a sauropod, who in particular is thought to have had high growth rates in the first several years after hatching to avoid predation. Unlike the extensive documentation for the blue whale, however, scant fossil evidence exists to help calculate how the largest dinosaurs grew. Limited fossil clues from relatively large but not the most massive dinosaurs permit a rough estimation of body growth rates, including relatively complete fossils of the duck-billed dinosaur *Maiasaura* at different stages of development. Measurements show that it grew from a hatchling length of 0.5 m to slightly more than half of its adult length of 7 m in its first year of life. That's about 3.6 m in the first year, or about 1 cm per day.

During this time, we know that *Maiasaura* had lines of arrested growth in its bones, indicating that it reached those 3.6 m in the first seven to nine months. If so, its daily growth rate would have been approximately 1.33 to 1.71 cm each day. Assuming its spinal axons grew at the same speed as well, that falls far short of the blue whale's peak axonal growth rate.

But wait! Perhaps the recent discovery of a juvenile sauropod, *Rapetosaurus krausei*, puts dinosaurs back in the competition. This new find provides the missing link between previously studied hatchlings and 15-meter-long adult specimens of this species. Comparing differences in the lengths of the femur between a hatchling and a juvenile *Rapetosaurus*, the extrapolated growth in body length is estimated to be up to 2.7 cm per day during the first months of life. Notably, this time span appears to be prior to the animal's first arrested growth period. As such, the average growth rate for the spinal cord axons matching body growth at 2.7 cm per day is competitive but still less than that for the blue whale, leaving the *Rapetosaurus* in a close second place.

Moving from examining full body length growth to a focus on axon growth in a rapidly extending neck could bring a new mammalian player into the competition. Prenatal giraffes have a relatively short neck, which is thought to protect them from injury during birth. Yet thereafter their neck can grow up to 2.5 cm each day. This suggests a similar rate of growth for their cervical spinal cord axons, which falls short of the blue whale and *Rapetosaurus* peak spinal axon growth rate. Still, a potential evolutionary mishap in the trajectory of a nerve in the neck may help the giraffe surge past the blue whale in the race for the most extreme axon stretch growth.

Following a direction that makes little anatomical sense, the giraffe's left vagus nerve exits the brain stem and travels down the neck, where it splits off into the left recurrent laryngeal nerve. That in turn loops under the aortic arch, part of the large artery carrying blood from the heart, and then travels back up the neck to the vocal cords. During development, the axons in the animal's left vagus and recurrent laryngeal nerve must grow roughly twice as fast as its neck does. This brings the growth rate of these axons to approximately 5 cm each day. So the giraffe appears to take the lead from the blue whale.

The Winner Is ...

Hold on, the *Rapetosaurus* also had a long neck, which is thought to have had the same odd anatomical distribution of the recurrent laryngeal nerve as the giraffe's neck, bringing the sauropod back into the contest. But with *Rapetosaurus*'s neck accounting for only about half the length of its body, even if we double the growth rate for the recurrent laryngeal nerve, we are back to approximately 2.7 cm per day. So in the final stretch, the winner is the giraffe–by a neck! At least that is the case for now. Other sauropod species were much larger than *Rapetosaurus*, and if they developed along the same time line, much more spectacular rates of axon growth may have occurred. That enticing possibility can only be explored by future research on the growth rates of the largest dinosaurs.

Humans may also soon enter the competition for the most extreme axon stretch growth. No, not by having people rapidly grow supersized necks and bodies, but by pushing the limits of axon growth in the laboratory. Indeed, extreme axon stretch growth of 1 cm a day has already been achieved in a lab dish using bioreactors that mimic nature by mechanically elongating axon tracts spanning two populations of neurons. Much faster rates of growth are thought to be possible as mechanisms of this otherwise mysterious form of axon growth are experimentally revealed. With this promising experimental work, as well as emerging understanding of rapid nervous system development in other species, the competition is far from over.

About the Authors

Douglas H. Smith is Robert A. Groff Endowed Professor of Neurosurgery and director of the Center for Brain Injury and Repair at the University of Pennsylvania.

Jean-Pierre Dollé is a senior research investigator at the department of neurosurgery and Center for Brain Injury and Repair at the University of Pennsylvania.

Mathew J. Wedel is an associate professor of paleontology at the College of Osteopathic Medicine of the Pacific and the College of Podiatric Medicine at the Western University of Health Sciences.

How Snakes Breathe while Crushing Prey

By Lars Fischer, Joanna Thompson

When boa constrictors and other strangling snakes wrap their prey in a deadly embrace, they don't just exert pressure on their victim; they put the squeeze on their own lungs as well. Now new research shows how these remarkable reptiles use a sophisticated breathing technique to avoid suffocating themselves.

Instead of using a diaphragm muscle to inflate their lungs as mammals do, snakes activate a series of muscles around their extremely long rib cage. But crushing a struggling animal restricts these muscles, and scientists have long puzzled over how the snakes survive this constriction contradiction.

A team of researchers, led by John Capano of Brown University, reports in the *Journal of Experimental Biology* that boa constrictors can selectively move individual rib muscles in whatever parts of their chest are unblocked at a given moment. This lets small areas of the lungs function like a pump, sucking air through the constricted zones to absorb as much oxygen as possible. "Oftentimes, when they're not ventilating with that [squeezed region of the lung], it'll be more or less entirely collapsed," Capano says.

To discover this process, the researchers wrapped boa constrictors with blood pressure cuffs to prevent certain parts of their chests from expanding. Then they measured respiratory flow using small masks strapped to the reptiles' snouts. Unfortunately, the snakes proved to be mask skeptics. "One of the big challenges was really just getting the animals to perform in the lab setting," Capano says.

But the researchers eventually saw how the snakes used selective breathing to take in adequate air despite the cuffs. Recordings of electrical activity and x-ray images confirmed that nerve impulses strategically activated specific muscles in free areas. The snakes did not even attempt to breathe with the restricted parts of their rib cage–instead they exclusively used muscles on ribs that could still move. The researchers suggest this ability developed early in the snakes' evolutionary history because it would have let them swallow ever larger prey.

"The authors made really clever biomechanics experiments," says Eletra de Souza, a snake researcher at the University of São Paulo, who was not involved in the study. "The bony and morphological adaptations of snakes always surprise me. It's amazing to see how these animals cope so well in the absence of limbs."

About the Authors

Lars Fischer is a chemist and works as a journalist and editor at Spektrum der Wissenschaft.

Joanna Thompson is an insect enthusiast and former Scientific American *intern. She is based in New York City. Follow Thompson on Twitter @jojofoshosho0.*

The Surprising Secret of Snakes' Venomous Bites

By Riley Black

The world hosts hundreds of wildly different venomous snake species, from brightly banded coral snakes to camouflaged cottonmouths. But somehow even distantly related species independently evolved specialized fangs with venom-carrying grooves–a longtime puzzle spurring new research into how this could have happened. The answer, says Flinders University herpetologist Alessandro Palci, has been hiding inside the serpents' mouths all along.

It turns out that the teeth of most snake species, Palci and his co-authors report in the *Proceedings of the Royal Society B*, have a ring of tiny indentations around their bases. This pitted tooth tissue, called plicidentine, helps to anchor the teeth to the jaws. "Before our study, plicidentine was thought to be limited to only three types of lizards among living reptiles," Palci says. In fact, the study finds, it seems to be ubiquitous in snakes.

And these dental anchors have taken on another function in venomous species, the researchers say. They used microscopic tissue sampling and micro CT scans of snake teeth to determine that in the teeth nearest a snake's venom glands, the plicidentine's folds have evolved into longer grooves that help snake venom flow from gland to tooth–and then into prey.

This result suggests that modern venomous snake species did not have to start from evolutionary scratch every time they developed a deadly bite. Many snakes produce toxic saliva, Palci notes, and a tooth with a more pronounced plicidentine groove lets that incipient venom travel more efficiently. Repeat this change in generations of snakes, and venom-delivering fangs will evolve.

"I love seeing modern imaging technology and beautiful micro CT scans applied to a classic question," says Whitman College herpetologist Kate Jackson, who was not involved in the new study. Snakes may have developed venom-delivering teeth via other routes as well, Jackson adds. For instance, many fish-eating species have grooved teeth that help to grip slippery prey; these grooves could also have evolved into venom channels. Nevertheless, she says, the study reveals a new common feature of snakes and can inform herpetologists' understanding of how snake fangs evolved anew time and again.

About the Author

Riley Black, who formerly wrote under the name Brian Switek, is the author of Skeleton Keys *and* My Beloved Brontosaurus. *She lives in Salt Lake City, Utah. Follow her on Twitter @Laelaps and on Instagram @laelaps.*

Genes Reveal How Some Rockfish Live up to 200 Years

By Jack Tamisiea

Few groups of animals encapsulate the extremes of longevity more than fish. While coral reef pygmy govies survive for less than ten weeks, Greenland sharks can endure more than 500 years. So when a team of biologists at the University of California, Berkeley, wanted to explore the genetics of aging, they grabbed their fishing gear.

Their preferred catch was rockfish. Found in coastal waters from California to Japan, rockfish are a colorful group of more than 120 species in the genus *Sebastes*. Some of these closely related species live for only a decade. Others, such as the rougheye rockfish, can live for more than 200 years.

The diversity of rockfish life spans offered the perfect parameters for parsing the genetics behind longevity, according to U.C. Berkeley biologist Peter Sudmant. In a study published on Thursday in *Science*, Sudmant and his colleagues sifted through the genomes of 88 rockfish species and pinpointed 137 specific genes that boosted the life spans of species such as the rougheye.

Some of the fishes' longevity comes from their size and habitat. When it comes to achieving longer life spans, scientists have found that bigger organisms often have an advantage because they have a slower metabolism and are less susceptible to predation. Similarly, colder environments can slow a creature's metabolism: for example, frigid waters make it possible for Greenland sharks to survive for centuries.

The study confirmed that some rockfish use both strategies by slowly growing large in the chilly waters along the seafloor as they spawn new generations. "Rockfish are the perfect storm: they live deep in the water, and some of these old moms grow to be very large," Sudmant says. "You have all these things at play that allow them to live for a very long time."

By comparing the genomes of short-lived rockfish with longer-lived species, the researchers were also able to identify particular genes that increase longevity. For example, longer-lived rockfish sported more genes associated with mending damaged DNA, presumably making them less susceptible to certain cancers than their shorter-lived kin. The researchers also found that longer-lived species had multiple genes geared toward regulating insulin, a capability linked to longevity in other animals. Another group of genes, known as the butyrophilins, regulates the rockfish immune system. Sudmant notes that similar genes have been tied to suppressing inflammation in aging humans.

While these genes' benefits may conjure thoughts of science-fiction-esque crossbreeding experiments, Sudmant is quick to pump the breaks. "We're not going to make hybrid rockfish humans and live forever," he says. This study's finding, however, "allows us to potentially think about what pathways and genes should be the key targets in our studies on humans and the drugs we develop."

Longevity may be an aspiration for humans, but it is a requirement for certain rockfish species, says Sabrina Beyer, a fish ecologist pursuing her Ph.D. at the University of California, Santa Cruz. It takes some species decades to reach reproductive maturity, and few of their larvae reach adulthood. To combat these long odds, female rockfish need to produce vast amounts of larvae per year. According to Beyer, who researches rockfish in affiliation with the National Oceanic and Atmospheric Administration's National Marine Fisheries Service and was not involved in this study, older females produce disproportionately more larvae than their younger counterparts. "There's a real incentive in these rockfishes to grow big and live long and produce lots of high-quality larvae," she says.

Many rockfish are commercially and recreationally fished for food along the West Coast, threatening certain species with extinction. Overfishing and environmental shifts prove more disastrous for long-lived rockfish species because they usually exhibit lower genetic diversity than their short-lived kin, Beyer says. "We can't have heavy fishing pressure on these long-lived, slow-growing species," she adds. "It's just not what they're adapted to."

About the Author

Jack Tamisiea is a science journalist based in Washington, D.C., who covers natural history and the environment. Follow Tamisiea on Twitter @jack_tamisiea.

Here's How a Python Jaw Can Fit a Whole Deer

By Bethany Brookshire

Invasive Burmese pythons slithering through the Florida Everglades eat almost anything they can get their jaws around–and that's a lot. As the snakes' numbers have risen, small mammal populations have plummeted. But larger animals aren't safe either; people have spotted these pythons swallowing alligators and white-tailed deer whole. How do snakes with mouths a few inches wide devour something that huge?

Answer: They open their mouths very wide, with the help of some newly measured stretching power. A recent study shows Burmese pythons' mouths can stretch four times wider than their skulls, creating a gaping maw four to six times bigger than a similarly sized brown tree snake's.

Most snakes cannot take bites and must engulf prey whole. To do so, a typical snake opens its mouth at the joint in the middle of its jaw, and the two halves of the lower jaw flare out to the sides; the skin and tissue in between stretch to accommodate the food. The skin eventually snaps back, but "after they swallow a very big meal, [their chins] are baggy for a while," says University of Cincinnati vertebrate morphologist Bruce Jayne.

And some snakes can open wider than others. For a study in *Integrative Organismal Biology*, Jayne and his colleagues examined Burmese pythons from Florida and brown tree snakes from Guam (where the latter are invasive). The researchers took anatomical measurements of snakes after death, then stretched the reptiles' jaws with funnels of increasing size. Finally, the scientists stuffed potential prey–including anesthetized alligators–through the funnels and measured deer remains recovered from a python stomach.

The experiment showed just how wide each species' mouth could get. The secret to the Burmese pythons' superior skill? Extra stretch in the tissues between their jaw bones. Some 43 percent of their gap width capacity could be attributed to this tissue, compared with 17 percent for the tree snakes.

According to Marion Segall, a herpetologist at London's Natural History Museum, who was not involved in the study, the two distantly related invasive snakes make a good comparison because they have each evolved to snag large prey relative to their sizes. Future work will explore what properties make the pythons' jaws so flexible.

Just because a python can eat a whole deer doesn't mean venison is often on the menu. "Most [tree snakes and pythons] are opportunistic and will catch anything that passes by, [so] they will probably not target the largest prey," Segall says. But to avoid presenting such an opportunity in the Everglades, even a big deer should probably sleep with one eye open.

About the Author

Bethany Brookshire is a freelance journalist who covers physiology, neuroscience and human-wildlife interactions. Follow her on Twitter @BeeBrookshire.

Section 3: Animals in Extreme Places

Deepest Fish Discovered More Than 5 Miles below the Sea Surface

By Tom Metcalfe

Scientists exploring a marine trench near Japan were astonished to see a fish in one of the deepest parts of the ocean, at 8,336 meters (about five miles) below the surface. The tadpole-shaped, translucent snailfish is probably living at the greatest depth possible.

"They can't really go any deeper," says marine scientist Alan Jamieson of the University of Western Australia, who led the team that made the discovery. The previous record holder, a snailfish seen in the Mariana Trench–the world's deepest location–was filmed 8,178 meters under the surface in 2017.

Fish can tolerate high pressures at extreme depths because of cellular compounds called osmolytes. Osmolyte concentrations increase at greater depths to ensure that fish cells can withstand such bone-crushing pressures, but these compounds reach their maximum concentration at around 8,400 meters. So that's the theoretical limit of fish physiology. "If anyone does find fish deeper than this, it will not be by much," Jamieson says.

Ichthyologist Prosanta Chakrabarty, curator of fishes at Louisiana State University's Museum of Natural Science, is impressed that the fish could survive at 800 times the surface's water pressure. "At that depth everything from gas exchange for breathing to nearly every physiological function seems impossible," he says. "I can barely swim to the bottom of a swimming pool without my ears popping."

Jamieson's team discovered the snailfish in August 2022 at the bottom of the Izu-Ogasawara Trench, near the main islands of Japan. The farthest depths of the Japanese trench reach about 1.7 degrees Celsius (35 degrees Fahrenheit), Jamieson says, which is slightly warmer than the neighboring Mariana. This divergence is key: osmolytes are less effective at low temperatures, and these snailfish live near the edge of what's possible. "The difference is a fraction of a degree, so we wouldn't care," Jamieson says. "But it makes a difference to marine animals."

To photograph the fish, researchers onboard the DSSV *Pressure Drop* sent down an autonomous underwater "lander" equipped with cameras, lights and batteries, along with a weight to lower the contraption to the seafloor. The lander carried dead fish as bait to lure deep-sea crustaceans, and the snailfish came to eat the crustaceans–including the record-breaking juvenile snailfish at 8,336 meters. Although the team couldn't identify its exact species, two others from the species *Pseudoliparis belyaevi* were caught in baited traps nearby, at a depth of 8,022 meters.

Each of the more than 400 known snailfish species adapts to where it lives, from shallow waters to extreme depths, Jamieson says. "Each trench has its own snailfish in it," he says. "Once they've evolved to cope in a trench, they cannot decompress to get from one trench to another."

About the Author

Tom Metcalfe is a freelance journalist who is based in London. Metcalfe writes mainly about science, space, archaeology, Earth and the oceans. He has also written for Live Science, *the BBC, NBC News,* National Geographic, Air & Space *and many others.*

Surprise Fossil-Munching Sponges Found in Arctic Seafloor Wasteland

By Jack Tamisiea

Entombed below a thick layer of sea ice year-round and nearly devoid of nutrients, the central Arctic Ocean is a frigid wasteland where few creatures are capable of surviving.

That is why researchers at Germany's Alfred Wegener Institute (AWI) and their colleagues were so surprised to spot a thriving community of velvety deep-sea sponges below the ice in 2016. Their ice-breaking research vessel was north of Greenland just around 200 miles from the North Pole, when the submersible camera they were towing caught sight of a garden of fuzzy sponges carpeting the tops of several extinct volcanoes like mold covering a carton of raspberries. Some of them even stretched more than three feet across–giants by deep-sea sponge standards.

The existence of so many large sponges was puzzling. Sponges, which lack a digestive tract, primarily rely on passive filter feeding to gather nutrients from water flowing through them. But the ocean currents in the central Arctic Ocean are sluggish, and few particles swirl in the water. Tests of the sponges' tissue also revealed that this colony was far from ephemeral–their average age was 300 years. That raised a question: How had these simple animals survived for centuries in a marine wasteland? "Sponges are very primitive animals," says Teresa Morganti, a biologist at the Max Planck Institute for Marine Microbiology in Bremen, Germany. "It's really interesting to think about how they can adapt to these extreme environmental conditions."

To search for secrets to the sponges' survival, the AWI researchers collected samples of the organisms and the sediment around them, which they sent to Morganti and her colleagues to examine. Their analysis revealed that the seafloor was not always barren: the sediment samples were full of fossils.

Many of the fossils were the hollow shells of large deep-sea worms that anchor themselves to the seafloor with pipelike tubes. Although no tube worms live there today, the scientists were not surprised to find their empty shells. Thousands of years ago, gases seeped from the hydrothermal vents of these submerged volcanoes, creating an ideal habitat for the worms.

That once dynamic seep ecosystem is still influencing the area long after the volcanoes sputtered out, the researchers concluded in a study published on Tuesday in *Nature Communications*. Samples the team collected suggest that the sponges were brimming with bacteria, including symbiotic microbes that appeared to help the animals acquire and process nutrients from the fossilized remains of their predecessors. "One community dies, and there is another organism–in this case, sponges–that colonized this area, and they use the ancient community not only as substrate but also as a food source," Morganti says.

Heidi Meyer, a biologist studying deep-sea sponge grounds at Norway's University of Bergen, was intrigued by the discovery that these deep-sea sponges were feeding on fossilized detritus. She believes this unprecedented feeding style opens up the possibility that there are more of these sponge oases throughout the Arctic Ocean. "I definitely believe that there are more sponge grounds that are similar to this positioned all along the [volcanic] ridge," says Meyer, who was not involved with the new study.

Such oases would be good news for a variety of other creatures. Like coral, sponges are effective ecosystem engineers. As they grow, they create a variety of nooks and crannies for other animals to live in. The sponges also deposit a sticky, matlike surface for bacteria and detritus to settle on. According to Morganti, this acts like an "organic matter trough" that attracts other deep-sea animals. When they surveyed the Arctic sponge grounds, the researchers also discovered tiny shrimp, small worms and pudgy sea stars. Even tufts of deep-sea coral poked out from the fossil-eating sponges below.

As the resourceful sponges build on the ruins of the ancient vent ecosystem, they create new ecological opportunities, Morganti says. "These sponge grounds are hotspots of life in the desert," she adds.

About the Author

Jack Tamisiea is a science journalist based in Washington, D.C., who covers natural history and the environment. Follow Tamisiea on Twitter @jack_tamisiea.

These Sharks Hold Their 'Breath' to Stay Warm

By Meghan Bartels

A dramatic-looking shark holds its breath for about 17 minutes to stay warm during hunting dives in cold water, scientists have found.

Scalloped hammerhead sharks (*Sphyrna lewini*) are found in warm, near-surface ocean waters around the world, particularly coastlines, and can reach up to around a dozen feet in length. But these sharks do venture into deep, chilly water to hunt. That's potentially problematic because scalloped hammerheads, like most sharks, are cold-blooded, which means their body temperature matches that of their immediate environment. "When you go into cold water, if your body cools down, you can't hunt as effectively," says Mark Royer, a shark biologist at the Hawaii Institute of Marine Biology. "In order to be an effective hunter, it's important to keep yourself warm as much as possible."

Some fish such as bluefin tuna, great white sharks (*Carcharodon carcharias*) and swordfish have special muscles and blood vessels that allow them to keep their brain, eyes and swimming muscles warm even in cold waters–but scalloped hammerheads don't. Yet as Royer and his colleagues analyzed 106 deep dives made by six individual scalloped hammerhead sharks, they saw that the animals stayed mysteriously warm long into these excursions, according to sensors that recorded the sharks' muscle temperatures, orientations, and diving depths and the ambient temperatures of the water around them.

"We looked at the body temperature data, [and] right off the bat, we could see that these sharks were doing something very interesting and very different, compared to any other fish," Royer says. "It's obviously not simple thermal inertia that these sharks are relying on to maintain their body temperature on these deep dives. There's something much more complex."

The researchers found that the sharks were making steep, fast descents to spend a few minutes feasting in deep waters, then speeding most of the way back to the surface. The animals stayed warm until the point where their ascent slowed—on average, 17 minutes, Royer says. The team's findings were published on May 11 in *Science*.

The researchers think the mystery lies in the sharks' gills. These organs act as lungs, pulling oxygen from seawater. But gills come with a downside because vast quantities of water pass through them. "Gills are like giant radiators strapped to your head," Royer says. "Whenever a shark or a fish goes into colder water, it's going to very quickly lose its body heat through its gills because the gills have all that large surface area."

The team couldn't monitor the sharks' gills directly, although Royer says that in the future, he'd like to equip sharks with pectoral-fin cameras pointing toward the gills. But for now, the scientists think that the sharks are closing their gills—holding their breath, essentially—during their daring dives. (Royer notes that other researchers have footage of one scalloped hammerhead diving with its gills shut.)

"It's definitely something that hasn't, as far as I know, been put forward for an ectothermal [cold-blooded] fish yet that I think is going to create really cool avenues for future research," says Ashley Stoehr, a marine biologist at Sacred Heart University, who specializes in fishes and wasn't involved in the new research.

Scalloped hammerhead sharks may not be unique in employing this technique for deep dives, Royer says. Oceanic whitetip sharks (*Carcharhinus longimanus*) don't make these dives as often as scalloped hammerheads do, he says, but they are still known for dramatic dives. And scientists know that these sharks also lack the specialized muscles and blood vessels that can keep fishes warm. Still, it seems scalloped hammerheads' breath-holding technique is a rare approach to daring cold waters. "It's an incredible and unexpected behavior from a remarkable species," Royer says.

About the Author

Meghan Bartels is a science journalist and news reporter for Scientific American *who is based in New York City.*

How the Woolly Bear Caterpillar Turns into a Popsicle to Survive the Winter

By Kate Furby

Kate Furby: Some caterpillars have evolved with antifreeze in their body cavities, allowing them to become cater-Popsicles to survive cold winters. But climate change could threaten that.

Martha Weiss: So there are caterpillars that have been reported to be put into an ice cube and frozen, and then when the ice cube melts, they can get up and walk away.

Furby: You may have seen them scooting around on leaf litter in the fall. They're furry, rotund, and famous for their rumored weather forecasting skills. I'm talking about the woolly bear caterpillar, or Isabella tiger moth.

These little creatures have an orange waistband stripe, whose width is rumored to predict how long winter might be. And while this is based in colonial folklore, not science, what *is* scientifically amazing is how the woolly bear caterpillar is able to survive winter.

I'm Kate Furby, and you're listening to *Science, Quickly*.

Furby: Unlike humans and other mammals, caterpillars can't regulate their body temperatures. And unless they burrow or cocoon, they're subject to the wind and rain. The woolly bear caterpillar, like its name, is covered in a spiky looking fuzz.

Weiss: Those hairs you might think of as a little down jacket for the caterpillar to wear, and I'm sure that they do provide a little bit of insulation.

Furby: That's Dr. Martha Weiss, a biologist and professor at Georgetown University, who studies plant-insect interactions. She says that spiky little caterpillar jacket has a specific use but not what you might think.

Weiss: Those hairs are thought to have evolved as a way to protect the caterpillars against predators and maybe against parasitoids that wanna lay their eggs inside the caterpillar's body.

Furby: Yikes, that is a superpowered little jacket actually. But here's the caterpillar's dilemma.

Weiss: Well, the main thing is that it gets really cold, and they have a lot of water in them, and they can freeze. And so they need to be able to deal with freezing temperatures.

Furby: And while the famous furry jacket provides protection, it doesn't provide the kind of insulation woolly bear caterpillars need for a hard Chicago winter. What they do is a little more biochemical.

Weiss: They have more biochemical tricks up their sleeves insofar as caterpillars could be said to have sleeves.

Furby: Oh wow, they would have to have like 16 little sleeves! But okay, so what are their options for survival?

Weiss: They can also do biochemical things and physiological things to make it less likely that they will turn into an ice cube. So what some of those caterpillars do is: they use antifreeze. They basically make compounds like glycerol that they put into their cells.

Furby: In case you're not familiar with glycerol, it's a natural alcohol compound. It works similarly to when we salt city sidewalks to keep them from becoming icy. The compounds in the woolly caterpillar's body lower its freezing point, buying it some time. And then they do something even more remarkable.

Weiss: They move water out of their cells so that it freezes in the extracellular space.

Furby: That's because ...

Weiss: Water of course gets bigger when it turns to ice. And so if a cell was filled with water, and it froze, then it could bust the cell membrane, and that would really harm the caterpillar. So getting the water out of the cell is a good idea, and lowering the temperature at which the liquid freezes is also a good idea.

Furby: So these little guys can freeze solid all winter and then thaw out and get up and walk away come spring.

Weiss: They can actually freeze and thaw multiple times over the course of a winter.

Furby: But there's an energetic cost that comes to falling asleep and waking back up again.

Weiss: Studies of the Isabella tiger moth have shown that they can in fact undergo several freeze thaw cycles, but it's really not great for them. It's better if they can freeze, stay frozen and then thaw at the end of the winter.

Furby: And not only that ...

Weiss: I think there's also some damage that happens to some of the structures in the caterpillar. Some of the more delicate parts, I think, can be damaged a little bit. And the more times they have to freeze, thaw and refreeze, the more likelihood that they'll be a little worse for wear at the end of the winter.

Furby: And this gets worse because of things like climate change.

Weiss: If we have winter heat waves or warm periods when caterpillars that had been in the deep freeze thaw out and then freeze again, there was some concern that they wouldn't be able to go back and forth between these conditions.

Furby: That'll also lead to larger ecological implications.

Weiss: Caterpillar populations and therefore butterfly or moth populations could take a hit if the overwintering survival is interfered with by these interludes of warmer weather that prevent them from getting through their wintering period in the same way that they had before.

Furby: And that might have more profound impacts than we think. We already know that some important pollinators, like bees and butterflies, are struggling to survive due to all kinds of human activities.

Weiss: Caterpillars are important just because they're such cool animals, but they also are a phase of the life cycle of lepidopteran, so moths and butterflies. They are pollinators; they're herbivores; they're food for birds and other organisms. And they're just part of what makes the world fun to look at and live in.

For *Scientific American*'s *Science, Quickly*, I'm Kate Furby.

About the Author

Kate Furby is a National Geographic Emerging Explorer and the founder of Tiny Beaker Media. She has a Ph.D. from Scripps Oceanography in marine biology, specializing in coral reefs and climate change.

No One Knows How the Biggest Animals on Earth—Baleen Whales—Find Their Food

By Kate Wong

When it's time to eat, humpback whales head toward the ends of the earth. Their mission: feast until they are fat and happy. They must build up their energy reserves, packing on nearly a ton of blubber a week to sustain them on the voyage from their polar and subpolar feeding grounds to the balmy waters where they breed. The journey may require traveling thousands of miles over several months—and they must be ready to reproduce when they arrive. Perhaps because nature loves a paradox, these colossal predators, which can measure 60 feet long and weigh 40 tons, accumulate these fat stores by eating some of the smallest prey in the sea—including krill, shrimplike crustaceans that live in all the world's oceans but are concentrated in the cold waters found at high latitudes.

We know a lot about how humpbacks eat. They filter seawater through plates of keratin, called baleen, that line their upper jaws and resemble the frayed bristles of a worn toothbrush. They devour several thousand pounds of their tiny prey every day. To obtain that quantity of food, they must seek out dense aggregations of the crustaceans. Once they find a swarm, they may deploy a clever cooperative hunting tactic, swimming in circles while blowing columns of bubbles to create a kind of net to corral the krill. Then they feed, lunging at the tightly gathered prey with jaws agape, engulfing thousands of gallons of krill-filled water in their pleated throat pouches before straining the catch through their baleen.

Yet for all scientists have learned about these charismatic leviathans, no one knows how baleen whales (a group that includes humpback, blue, fin and sei whales, among others) find their food in the first place. Their cousins the toothed whales—sperm whales, belugas, dolphins, and the like—use ultrasonic sonar signals to detect prey, but baleen whales don't have that ability. Somehow they still manage to find their minuscule quarry in the infinite sameness of the sea.

It's a mystery that scientists are eager to solve. In part that's because it is a huge gap in our basic knowledge of high-profile species. More urgently, the question of how baleen whales seek out their food has important conservation implications, particularly for a baleen species called the North Atlantic right whale.

The North Atlantic right whale, a dark, stocky cetacean that eats rice-size zooplankton called copepods, has the unfortunate distinction of being one of the most endangered mammals on the planet. Commercial whaling nearly extinguished this species in the early 1900s. By 1935 the League of Nations banned the hunting of all right whales. But unlike other species whose numbers plummeted because of whaling, the North Atlantic right whale has been unable to make a comeback. The animal's feeding grounds off the coast of New England and the Canadian Maritimes overlap with areas of intense human activity. Collisions with ships and entanglement in fishing gear, along with climate change–induced disturbance of their habitat and prey, have taken a terrible toll.

The most recent estimates indicate that fewer than 350 North Atlantic right whales remain, only 70 of which are females of reproductive age. According to some projections, the species could go extinct in the next couple of decades. Understanding how baleen whales track down their prey could help scientists predict where the whales will go to feed—and better manage human activities in those areas that might harm the whales.

All of this matters for more than just a single species of whale. North Atlantic right whales and other baleen whales are ecosystem engineers, feeding in deep water and then releasing nutrients near the surface through their feces, which support the growth of microscopic plantlike organisms called phytoplankton. The phytoplankton, in turn, nourish krill, copepods and other tiny drifting creatures known as zooplankton that are eaten by larger animals. The whales' tissues also trap enormous amounts of carbon dioxide that could otherwise contribute to global warming–an estimated 33 tons for the average large-bodied whale. And when whales die, their carcasses sink to the seafloor, where they sustain entire communities of deepwater organisms–from sleeper sharks to sulfur-loving bacteria–that are specially adapted to using these so-called whale falls for food and shelter. The health of baleen whale populations supports the health of a host of other species.

The most direct way to learn how a baleen whale finds its food is to tag it with a device that can record its underwater behavior and watch the animal forage. That's not possible with North Atlantic right whales, which are so stressed from human activity that any direct human contact could just make things worse. Fortunately, the right whale has cousins, such as the humpback, that are in much less peril. And one of the best places to watch them eat is on their feeding grounds at the bottom of the world.

In 2020, two weeks before the World Health Organization declared the COVID-19 outbreak a pandemic, I boarded a ship to Antarctica to follow one research group's efforts to learn how baleen whales find food. I went as a guest of the cruise operator, Polar Latitudes, to observe a study being carried out by the seven scientists they were hosting on their tourist boat and to lecture on whale evolution.

By joining a tourist expedition, the international team of researchers based in the U.S., Sweden and Japan saved on the exorbitant costs of getting to the white continent. In return for three shared staterooms, meals, and the use of two sturdy inflatable rubber boats called Zodiacs, the scientists gave the other passengers regular updates on their research, which was billed as a whale-focused expedition for citizen scientists.

The team was testing a hypothesis about baleen whale foraging that grew out of research on seabirds. Starting in the mid-1990s, Gabrielle Nevitt of the University of California, Davis, showed that dimethyl sulfide (DMS), a chemical that is released when phytoplankton are eaten by zooplankton, attracts tube-nosed seabirds–a group of carnivorous birds that includes albatrosses, petrels and shearwaters–which then eat the grazing zooplankton. It's a mutualistic arrangement: by luring the seabirds with the scent of DMS, the phytoplankton gain protection from the zooplankton. Even at the bottom of the food chain, the enemy of your enemy is your friend.

Cruise team leaders Daniel Zitterbart of the Woods Hole Oceanographic Institution, a physicist who uses remote sensing methods to study the behavior and ecology of whales and penguins, and whale behavior specialist Kylie Owen of the Swedish Museum of Natural History wondered whether whales might be similarly attracted to DMS. If so, then following the chemical toward higher concentrations should, in theory, lead whales to denser concentrations of the krill and other phytoplankton eaters than foraging randomly would. To find out, Zitterbart and Owen joined forces with whale biologist Annette Bombosch of Woods Hole; zooplankton researcher Joseph Warren of Stony Brook University; Kei Toda of Kumamoto University in Japan, who developed technology for measuring DMS, and his then graduate student Kentaro Saeki; and oceanographer Alessandro Bocconcelli of Woods Hole, who has helped pioneer the use of sophisticated digital tags to study whales.

The team planned to tag humpbacks with custom-built instruments containing pressure sensors, accelerometers, magnetic compasses and hydrophones that record their underwater behavior, along with a radio transmitter to enable tracking. Their permits allowed them to tag only a total of five whales, and they had to do it in just five days–the rest of the 12-day cruise would be spent in transit. They had little room for error.

We left from the Argentine port of Ushuaia, the southernmost city in South America, on February 28 and spent the next two days of the leap year crossing the Drake Passage, the notoriously turbulent 620-mile-wide waterway between South America and Antarctica, escorted by albatrosses and petrels. On March 1, we passed over a boundary zone known as the Antarctic Convergence and entered the calm, cold waters of the Southern Ocean. For the first time since entering the Drake, we glimpsed land off the starboard side of the ship–Smith Island, part of the South Shetland Islands of the British Antarctic Territory.

With the stomach-churning swells of the Drake behind us and the soporific effects of the motion sickness medication wearing off, I could now fully register my extraordinary surroundings. Icebergs, bergy bits and growlers–some of the many forms of ice here–joined sea and sky to display every shade of blue. Fuzzy Gentoo penguin chicks chased after their exhausted parents to demand food. Platinum blond crabeater seals lounged on divans of drifting ice, basking in the sun. I let the otherworldly beauty of the place wash over me.

On the morning of March 4, I awoke to daybreak in Paradise Bay, a scenic harbor where whaling ships once anchored. From my seat on the pontoon of a Zodiac, I watched the rising sun pierce through an opening in the cloud cover to bathe a distant glacier in golden light.

We were in whale country now, encountering groups of the mammals as they floated on the surface like logs, exhaling tall plumes of moist air. The wet *whoosh* of whale breath joined the thunder cracks of calving glaciers and the rumbles of avalanches.

The day before, the scientists had successfully tagged their first humpback. The passengers cheered when the scientists announced the update at breakfast. Unfortunately, the whale proceeded to sleep the entire time it was under observation. But later the same day they tagged a second whale, and this one was a model subject, making several dives up to 850 feet. Data from the sensors indicate that the whale was lunge feeding–exactly what they wanted to see.

This morning the team was attempting to tag a third individual–and hoping it behaved like number two. Zitterbart, a tall, animated man who thinks and talks with formidable speed, got up at 5:30 and headed to the ship's bridge to find out whether any whales were around and what the weather was like. The day looked promising. Whales had been spotted in the area, and the water was still–the better for retrieving tags, which are programmed to stay on a whale for just a few hours before detaching and floating to the surface.

By 6:45 the research boats were lowered into the water, and the scientists were preparing to tag a whale that had been sighted nearby. A 20-foot-long carbon-fiber pole extended beyond the bow and stern of the tag boat. They use the pole to slap the tag, which has four suction cups on its underside, onto unsuspecting humpbacks once they get to within 10 feet of the animal. Bombosch and Bocconcelli navigated across a glassy expanse of open water toward a group of whales, slowing on the approach. That bunch looked lazy, though. They didn't want to tag another slumbering whale, so Owen and Bombosch decided to target another group that looked more active.

From my vantage point in a separate Zodiac, two humpbacks came into view. Only their small dorsal fins and the uppermost part of their sleek black backs were visible. They didn't look all that big. But like icebergs, most of their mass is below the waterline. At a distance, you only get a sense of how huge humpbacks are when they wave their great flippers in the air, raise their tail flukes ahead of a deep dive, or propel their entire bodies clear out of the water in a glorious breach.

Zitterbart gripped the unwieldy tagging pole and stood tensed, one foot on the bow box and one in the boat. Attaching the tag is a fraught operation. To ensure a strong signal from the transmitter, he had to place the tag as high on the animal's back as possible but not too close to the sensitive skin surrounding the blowhole. As the tag boat neared the whales, Zitterbart raised the pole and then, at exactly the right moment, cast it down with just enough force to plunk the tag securely onto one of the animals. The whale startled, then sank out of sight—a typical reaction—and the researchers moved quickly to stow the pole, mark the GPS location of the tagged whale and prepare to monitor the animal. They were now three for three with attaching the tags.

Once the tagged whale resurfaced, they would spend the next few hours tracking it by eye and with the aid of a VHF receiver tuned to the tag's transmitter, keeping a distance of more than 300 feet from the animal so as not to interfere with its routine. They needed to recover the tags—which store the behavior data and cost

$10,000 apiece–when they automatically detached from the whales at the preprogrammed time after deployment. Now the team just had to hope they chose a cooperative subject. "Ideally we'd tag an active whale that's not feeding yet, and it would swim away to feed," Owen explained. After that, the researchers in the prey boat would sample the water to see if krill and DMS concentrations were increasing along the whale's path. If they tagged a whale while it was already feeding, they would have no trail to follow. But the humpbacks are wild animals, with agendas of their own. "The stars really need to align for things to go the way we want them, too," Owen said.

To visit Antarctica is to encounter forces that have shaped the fortunes of baleen whales across eons. Descended from four-legged land animals, whales underwent one of the most dramatic transformations of any vertebrate group when they transitioned to life in the water. Like all organisms, whales evolved under the influence of environmental change. They got their start some 50 million years ago in the greenhouse conditions of the Eocene epoch. Back then, the southern supercontinent of Gondwana was in the process of disbanding, and the ancient Tethys Sea reached from the Pacific Ocean to the Mediterranean. In the warm, shallow waters of Tethys, early whales underwent the first phase of their transformation: becoming seaworthy. Forelimbs morphed into flippers, noses became blowholes, ears remodeled to hear while submerged. Some 10 million years after their furry, four-legged ancestors walked along the water's edge, whales had adapted so thoroughly to aquatic life that they could no longer venture ashore.

The second phase of whale evolution unfolded as the planet transformed into a so-called icehouse world. As the Eocene gave way to the Oligocene, tectonic forces dealt a final blow to Gondwana, cleaving apart Australia, South America and Antarctica. When the separation of these landmasses was complete, the Antarctic circumpolar current swept around Antarctica, isolating it from warmer waters and pulling up nutrients from the deep that supported an abundance of phytoplankton and zooplankton. So vast and powerful was this new current, in fact, that it altered ocean circulation, temperature and productivity across the globe. From this crucible of tectonic, climatic and oceanic change, the forerunners of modern baleen whales emerged. By 35 million years ago early representatives of this lineage were patrolling the seas. Over millions of years their descendants would eventually acquire the baleen and gigantic body sizes for which this branch of the whale family is known.

Although baleen whales were molded by dramatic environmental and ecological change on evolutionary timescales, that long history did not inoculate their modern descendants against the dangers of profound change on shorter timescales. In the 20th century alone, industrial whalers armed with exploding harpoons and factory ships that could process carcasses offshore slaughtered more than two million baleen whales, pushing many populations to near extinction and degrading their ecosystems. Some species have been recovering since the demise of that industry–only to now face a new round of existential threats. Warming seas and commercial fishing are changing the availability of the zooplankton the whales depend on for food.

Four days after observing the tagging operation, I joined Warren, Zitterbart, Saeki and Julien Bonnel of Woods Hole on the prey boat. The cruise ship had to make a detour to Frei Station, a Chilean base with an airstrip on King George Island in the South Shetlands, to evacuate an injured passenger to the nearest hospital, in Chile. The researchers decided to use the unexpected stop to map the krill and DMS concentrations in a shallow embayment on the island's north side.

We wore jackets, hats and gloves against the morning cold, but just a few weeks earlier Antarctica had logged an all-time high temperature of 64.94 degrees Fahrenheit. The Antarctic Peninsula, where we had been exploring, is one of the fastest-warming regions on the planet. As a result, it's losing large amounts of ice, which is bad for krill, Warren said. Juvenile krill depend on winter sea ice for shelter and are thought to eat algae that grow on the underside of the ice.

Rising temperatures are not the only source of pressure on krill. Demand for the small crustaceans has surged over the past two decades, mostly from the nutritional supplements industry, which promotes krill oil as a rich source of omega-3 fatty acids for humans, and the aquaculture industry, which uses krill in feed for farmed fish. Whether the krill fishery is being managed sustainably is a contentious question. But a 2020 study of krill predators found that even with conservative catch limits for Antarctic krill in the waters around the Antarctic Peninsula–less than 1 percent of the stock in the southwestern Atlantic sector of the Southern Ocean–penguins in this region are declining, perhaps because the fishing vessels are focusing their efforts in areas that the penguins also favor. As the distribution and biomass of krill and other prey species change, predators–including whales–have to adapt their foraging routines accordingly.

As the Zodiac chugged away from the cruise ship, the researchers set up their equipment. They use an echo sounder to send sound waves down into the water, where they bounce off krill and any other animals they encounter, generating a picture on Warren's laptop of the creatures drifting in the water column. The lower the frequency of the ping, the deeper the transducer can "see." Higher-frequency pings, in contrast, can see smaller targets. The team uses two frequencies, one low and one high, to search for aggregations of the tiny krill, which typically hang out in the upper 650 feet of the water column. Warren's lab mascot, a small squeaky-toy pig named Sir Pings-a-Lot II, was overseeing the proceedings. "This is as exciting as it's going to get," Warren joked as he dropped the echo sounder overboard.

Krill are not as thrilling to track as their predators, but in recent years the science that happens in the prey boat has produced the greatest gains. As the boat traveled along its transect lines, Saeki reached over the side to scoop seawater from the surface every two minutes for analysis. Two plastic cases about the size of a handbag and carry-on suitcase contained the equipment for measuring any DMS in the water samples. A bubbler pushes air into the sample to get the DMS into the gas phase; a dryer removes any lingering moisture; an ozonator creates elementary sulfur from the DMS gas; and a photomultiplier measures light emitted by the sulfur–the amount of light is proportional to the amount of DMS present. Previously this sort of analysis was done in the lab; Woods Hole researchers were able to miniaturize Toda's DMS measurement setup to fit into a small boat. "The fact that we can run the DMS sniffer in the Zodiac is the big accomplishment this season," Zitterbart said. Among other things, it allows them to analyze a water sample on the spot. "We don't know how long the DMS signal from the water sample is viable," he explained. "To be cautious, we process it within two minutes."

Monitor the echo-sounder data, scoop the water, process the sample. Repeat. There were no whales here to distract from the monotony, just achingly blue sky, a raw wind and the drone of the outboard. We were more than halfway through the survey before the echo sounder detected any krill–a patch of the crustaceans suspended above the seafloor in the shallow waters of the embayment. What the work lacked in adrenaline it made up for in potential scientific impact. "Nobody has surveyed much of these bays, so any data we can get are valuable," Warren said. They returned to the ship with two krill patches detected and dozens of water samples analyzed–data that will help researchers understand how krill and DMS are distributed in the Southern Ocean and establish a baseline for measuring future change.

By March the brief austral summer was already drawing to a close. Daylight was ceding time to darkness, and the sea ice was starting to advance. Soon the humpbacks would head north to breed in the warm waters off the western coasts of South and Central America. Maybe that's why they weren't cooperating. Although the researchers had successfully tagged the five whales they had permits for, only two of the creatures went on to feed while they were being monitored. The other three snoozed or milled around the bays relaxing. To Zitterbart, the whales' lack of interest in foraging means that next time the team needs to shift the timing of their research.

"By March [the humpbacks] are already so big that they're sleeping too much," he said. "Earlier in the season is better because the whales are still building their body reserves and are more active."

The water chemistry strategy may need tweaking, too. Preliminary analysis of the samples obtained by the researchers as well as additional samples gathered by passengers through the ship's citizen science program showed lower than expected signals from the DMS. Perhaps there just wasn't a lot of DMS in the water. But another possibility, Warren speculated, is that a layer of melted freshwater atop the seawater diluted the signal. "The physics of water complicates things," he said. To get a clearer picture of the chemistry, the researchers may need to sample deeper water.

Going forward, Zitterbart wants to move away from cruise-ship sightseeing schedules and focus on building a detailed picture of the activity in a single bay. The plan is to hitch a ride on a cruise to one of Antarctica's research bases and stay there with just the Zodiacs. They'd map the whales, the krill and the water chemistry in the same place multiple days in a row and see how they change, then catch the mother ship on its way back.

First, though, they need to find a boat that can take them back to the bottom of the world. The cruise industry has a backlog of paying customers from the past few years who were unable to go on their planned voyages because of the pandemic. Trips the team might ordinarily be able to tag along on are fully booked. "We anticipated needing five years of data, and now three years are gone," Zitterbart said of the pandemic's effect on the project. He's hopeful they may be able to get passage in 2024. In the meantime, he has turned his attention to research on the other side of the planet that could hasten help for the whales that need it most.

During the past three years, while waiting for the next Antarctic opportunity, Zitterbart, Owen and their colleagues have been studying the relationship among DMS, zooplankton and baleen whales in the waters off Massachusetts. Because they can't tag North Atlantic right whales, they're looking for correlations between DMS hotspots and right whale aggregations in Cape Cod Bay. The idea is to see if the chemical can be used as a proxy to predict where the whales will show up. The team surveys the whales by boat and plane, no tags required. Whereas the Antarctic research aims to identify the precise mechanism by which the baleen whales find their prey—whether it's by following DMS gradients to swarms of krill or some other means—the Cape Cod work seeks only to establish whether these whales tend to show up in parts of the ocean where DMS concentrations are higher. If so, then regardless of whether the whales are actually detecting DMS or following some other cue that just happens to be linked to DMS, the scientists can theoretically use DMS values to predict where and when whales will appear.

Current efforts to protect North Atlantic right whales involve seasonal speed restrictions for ships and visual and acoustic monitoring systems. For example, from January 1 to May 15 in Cape Cod Bay, an important feeding ground for the right whales, all vessels 65 feet long or longer have a speed limit of 10 knots to reduce the likelihood of serious injuries to whales from collisions. If whales are seen or heard in the area at any time of year, then boats of all sizes are asked to slow down and watch out for the creatures. A free app called Whale Alert displays seasonal management areas and whale-detection data on a map in near-real time.

But these management approaches lack predictive power, says research ecologist David Wiley of the National Oceanic and Atmospheric Administration's Stellwagen Bank National Marine Sanctuary, who works with Zitterbart on the DMS research. And big boats in congested shipping lanes often can't change course fast enough to avoid collisions with slow-moving whales. "With a predictive tool like DMS, we can plan rather than react."

In 2021 Owen, Zitterbart, Wiley and their collaborators published a paper based on the Cape Cod research showing that higher levels of DMS correspond to higher concentrations of zooplankton, so if baleen whales do track DMS, it will, in fact, lead them to prey. Now the researchers are looking at whether baleen whales actually aggregate in these DMS hotspots. Preliminary results indicate that both North Atlantic right whales and sei whales (another baleen species that eats copepods) do.

To strengthen their case, starting this year, the researchers will measure DMS concentrations in Cape Cod Bay and Massachusetts Bay every two weeks along standardized track lines before the right whales get there, when they arrive and when they leave. Their goal is to figure out how much DMS has to be in the water for the whales to show up. "We need to find out the thresholds, what's biologically relevant to the whales," Wiley says of the study, which he estimates will take around two years.

The dream is to be able to monitor places where DMS levels are building—and thus likely to be gathering spots for North Atlantic right whales—from space using satellite imaging. Wildlife managers could reroute ships around those areas or temporarily shut down fisheries or wind energy sites that might disturb the whales until the DMS levels subside and the whales move on. Climate scientists have long been interested in DMS because it promotes cloud formation. They have already found that the chemical can be detected from space. But it will take higher-resolution satellite data than is now available to predict the movements of whales.

For North Atlantic right whales and all the organisms whose fates intertwine with theirs, insights can't come fast enough. "If things don't change, right whales will go extinct in our lifetime," Wiley says. He believes the plight of this keystone species is the conservation issue of our time. Maybe with the help of hungry humpbacks in Antarctica and some curious scientists, North Atlantic right whales and other imperiled baleen whales will one day reclaim their place as rulers of the ocean realm.

About the Author

Kate Wong is a senior editor for evolution and ecology at Scientific American. *Follow her on Twitter @katewong.*

Section 4: Strange Connections Between Humans and Animals

Why Has a Group of Orcas Suddenly Started Attacking Boats?

By Stephanie Pappas

A trio of orcas attacked a boat in the Strait of Gibraltar earlier this month, damaging it so badly that it sank soon afterward.

The May 4 incident was the third time killer whales (*Orcinus orca*) have sunk a vessel off the coasts of Portugal and Spain in the past three years. The subpopulation of orcas in this region began harassing boats, most often by biting at their rudder, in 2020. Almost 20 percent of these attacks caused enough damage to disable the vessels, says Alfredo López, an orca researcher at the Atlantic Orca Working Group (GTOA), which monitors the Iberian killer whale population. "It is a rare behavior that has only been detected in this part of the world," he says.

Researchers aren't sure why the orcas are going after the watercraft. There are two hypotheses, according to López. One is that the killer whales have invented a new fad, something that subpopulations of these members of the dolphin family are known to do. Much as in humans, orca fads are often spearheaded by juveniles, López says. Alternatively, the attacks may be a response to a bad past experience involving a boat.

The first known incident occurred in May 2020 in the Strait of Gibraltar, an area with heavy boat traffic. Since then GTOA has recorded 505 cases of orcas reacting to boats. Sometimes they simply approached the vessels, and only a fraction of cases involved physical contact, López says. In a study published in June 2022 in *Marine Mammal Science*, he and his colleagues cataloged 49 instances of orca-boat contact in 2020 alone. The vast majority of the attacks were on sailboats or catamarans, with a handful involving fishing boats and motorboats. The average length of the vessels was 12 meters (39 feet). For comparison, a full-grown orca can be 9.2 meters (30 feet) long.

The researchers found that the orcas preferentially attack the boats' rudder, sometimes scraping the hull with their teeth. Such attacks often snap the rudder, leaving the boat unable to navigate. In three cases, the animals damaged a boat so badly that it sank: In July 2022 they sank a sailboat with five people onboard. In November 2022 they caused a sailboat carrying four to go down. And finally, in this month's attack, the Swiss sailing yacht *Champagne* had to be abandoned, and the vessel sank while it was towed to shore. In all cases, the people onboard were rescued safely.

In 2020 researchers observed nine different individual killer whales attacking boats; it's unclear if others have since joined in. The attacks tended to come from two separate groups: a trio of juveniles occasionally joined by a fourth and a mixed-aged group consisting of an adult female named White Gladis, two of her young offspring and two of her sisters. Because White Gladis was the only adult involved in the initial incidents, the researchers speculate that she may have become entangled in a fishing line at some point, giving her a bad association with boats. Other adult orcas in the region have injuries consistent with boat collisions or entanglement, López says. "All this has to make us reflect on the fact that human activities, even in an indirect way, are at the origin of this behavior," he says.

The safe rescue of everyone involved, however, suggests to Deborah Giles that these orcas don't have malevolent motivations against humans. Giles, science and research director of the Washington State–based nonprofit conservation organization Wild Orca, points out that humans relentlessly harassed killer whales off the coasts of Washington and Oregon in the 1960s and 1970s, capturing young orcas and taking them away for display at marine parks. "These are animals that, every single one of them, had been captured at one point or another–most whales multiple times. And these are whales that saw their babies being taken away from them and put on trucks and driven away, never to be seen again," Giles says. "And yet these whales never attacked boats, never attacked humans."

Though it's possible that the orcas around the Iberian Peninsula could be reacting to a bad experience with a boat, Giles says, it's pure speculation to attribute that motivation to the animals. The behavior does seem to be learned, she says, but could simply be a fad without much rhyme or reason–to the human mind, anyway. Famously, some members of the Southern Resident orcas that cruise Washington's Puget Sound each summer and fall spent the summer of 1987 wearing dead salmon on their head. There was no apparent reason for salmon hats to come in vogue in orca circles, but the behavior spread and persisted for a few months before disappearing again. "We're not going to know what's happening with this population," Giles says, referring to the Iberian orcas.

The Iberian orca attacks typically last less than 30 minutes, but they can sometimes go on for up to two hours, according to the 2022 study. In the case of the *Champagne*, two juvenile killer whales went after the rudder while an adult repeatedly rammed the boat, crew members told the German magazine *Yacht*. The attack lasted 90 minutes.

The Iberian orca subpopulation is considered critically endangered, with only 39 animals the last time a full census was conducted in 2011. A 2014 study found that this subpopulation follows the migration of their key prey, Atlantic bluefin tuna–a route that puts them in close contact with human fishing, military activities and recreational boating. Maritime authorities recommend that boaters in the area slow down and try to stay away from orcas, López says, but there is no guaranteed way to avoid the animals. He and his colleagues fear the boat attacks will come back and bite the orcas, either because boaters will lash out or because the attacks are dangerous to the animals themselves. "They run a great risk of getting hurt," López says.

About the Author

Stephanie Pappas is a freelance science journalist. She is based in Denver, Colo.

Gorillas' Resilience after Early-Life Trauma Holds Lessons for Humans

By Rachel Nuwer

Whether you're a human or an elephant, a baboon or a fish, adversity experienced early in life is often linked to negative effects on longevity and health in adulthood. But this tendency, seen across the animal kingdom, seems to have at least one exception: mountain gorillas. So long as young gorillas who experience adversity make it past the age of six, they will go on to lead lives just as long as their untraumatized peers, researchers reported on May 15 in *Current Biology*.

"For reasons we're still trying to understand, gorillas seem to be remarkably resilient to early-life adversity," says Stacy Rosenbaum, a biological anthropologist at the University of Michigan and senior author of the new paper. Understanding how one of our closest relatives copes with early life trauma could strengthen insights into why some people manage to do the same, she adds, whereas others continue to be haunted by negative childhood experiences throughout life.

Studying how long-lived animals in the wild respond to early-life adversity is difficult, and until now, researchers did not have much insight into the impact such events have on nonhuman great apes. Evidence suggests that chimpanzees and bonobos who suffer early-life traumas in captivity oftentimes go on to develop long-term psychological problems. But it's difficult to disentangle those observations from the fact that these animals are kept in unnatural settings.

The new study was made possible by the existence of a unique database currently managed by the Dian Fossey Gorilla Fund, a nonprofit conservation group in Rwanda. Launched in 1967, the database contains daily, lifelong observations of more than 400 mountain gorillas monitored in Rwanda's Volcanoes National

Park. "It's really, really unusual to have that kind of detailed demographic data to do a study like this, especially in wild apes," Rosenbaum says. "Literally hundreds of people over 55 years went out every day to watch the gorillas."

Rosenbaum and her colleagues focused on 253 gorillas from 19 social groups. They examined whether each animal experienced any of six types of early-life adversity before age six, including losing their mom or dad, living through group instability or witnessing the infanticide of a fellow young animal. Then the team looked at how long each of the animals in the study lived.

Young gorillas who were exposed to multiple adversities suffered up to nearly 30 times the risk of death before age six, the researchers found. But to their surprise, those individuals that did manage to survive past age six experienced no negative impact on how long they lived. In fact, males who experienced the highest levels of adversity tended to live longer than males who experienced the least amount of adversity—a finding that Rosenbaum and her colleagues think is likely attributable to viability selection. "The basic idea is that you only survive that early stuff if you're a particularly high-quality animal," she says.

For female gorillas, there was no difference in longevity, whether or not the animals went through a tough first few years. Taken together with the finding about male gorillas, this suggests that the species in general is "just more resilient," Rosenbaum says.

"It is quite surprising to learn that individuals who weather adversity in their early years do not suffer long-term consequences for survival," says Melissa Emery Thompson, an evolutionary anthropologist at the University of New Mexico, who was not involved in the new study. "Obviously, humans are not gorillas, but it is incredibly valuable to have comparative data on this phenomenon in a long-lived species and one that shares so many biological similarities with humans."

Rosenbaum and her colleagues do not have a definitive explanation for their findings, but they do have a hypothesis. For one, the gorillas whose lives are recorded in the database "live in this weird little bubble" where they do not face ecological scarcity, Rosenbaum says. There's plenty of water and food, and the forest is protected from loggers and poachers, so their physical needs are met. The animals also live in tightly knit social groups with high degrees of support. If a juvenile gorilla is orphaned, for example, other group members, particularly the alpha male, will step in to provide more care.

If this two-pronged explanation proves accurate, then it means that for social animals, evading the deleterious impacts of early trauma requires both social support and resources, Rosenbaum says. "Gorillas have both," she adds.

Many questions remain, including whether early-life adversity causes any impacts to gorillas' health and also what biological mechanisms underpin resilience in the great apes. Using the same database, Rosenbaum and her colleagues have already started working on a follow-up study to investigate health differences between traumatized and untraumatized gorillas as measured through things such as hormone levels obtained through fecal samples, records of illness and injury and relationship dynamics.

Despite the significant differences between humans and our closest great ape relatives, Rosenbaum emphasizes that "we do have a tremendous amount of things in common." She suspects that the essence of the gorilla findings probably applies to people and other long-lived mammals, too. "When we're thinking about how to help kids recover from early adversity, it's very important to keep in mind that it's not enough to just provide them with social support or to just provide them with economic support," she says. "If we ignore one or the other, I think the chances of success when doing interventions is not very high."

About the Author

Rachel Nuwer is a freelance science journalist and author who regularly contributes to Scientific American, *the* New York Times *and* National Geographic, *among other publications. Follow Nuwer on Twitter @RachelNuwer.*

Parrot Babies Babble Just like Us

By Karen Hopkin

Karen Hopkin: This is *Scientific American*'s *Science, Quickly*. I'm Karen Hopkin.

Little kids say the most nonsensical things, like my son at the aquarium when he was two or three.

[CLIP: Toddler talking about fish]

Toddler: "They say bloop bloop. I think that one and this one are saying bloop bloop."

Hopkin: Even sweeter are the sounds they make when they're just beginning to explore their vocal capabilities.

[CLIP: Baby babbling]

Hopkin: But human babies are not unique–at least when it comes to this sort of inarticulate articulation–because a new study shows that baby parrots also babble before they leave the nest. The findings appear in the journal *Proceedings of the Royal Society B: Biological Sciences.*

Karl Berg: When baby humans are babbling, they're kind of stringing a bunch of sounds together in sort of unintelligible ways. There's no identifiable context.

[CLIP: Baby human babbling]

Hopkin: Karl Berg is an associate professor of biology at the University of Texas Rio Grande Valley. He says parrot babble is really very similar.

Berg: Gibberish is a good descriptor.

[CLIP: Parrotlet babbling]

Berg: They're kind of mixing everything together, and it seems like they're regurgitating everything they've ever heard before.

Hopkin: Berg and his colleagues have been studying a species of parrot called the Green-rumped Parrotlet in Venezuela.

Berg: We've been intensively monitoring the population. We're in our 35th consecutive year this year.

Hopkin: The researchers positioned video cameras inside the birds' nests. But it took them a while to notice the baby parrots running through their vocal warmups.

Berg: We initially missed it because when we first started filming inside the nest cavities, all of the action and all the interesting things seemed to be happening when the parents would arrive.

Hopkin: That prompts the babies to start clamoring for food. But when their parents are away, the nestlings take turns napping, preening each other and talking to themselves.

Berg: Oftentimes the other siblings are asleep.

Hopkin: And they keep up this peeping and cheeping even after those siblings have flown the coop, leaving them the last little parrotlet left in the nest.

Berg: So they're spending endless hours in there alone, and they do look kind of bored, and we see babbling then as well.

Hopkin: This nestling nattering differs from adult conversation in that it's a little lackadaisical–and very, very quiet.

[CLIP: Parrotlet babbling]

Berg: If you're standing just a few feet from the nest, I suspect you would not hear it.

Hopkin: Whereas mature calls are much faster paced ...

[CLIP: Mature parrotlet call]

Hopkin: And generally more strident.

Berg: Parrots are kind of known for being noisy and loud.

[CLIP: Pair of parrotlets dueting]

Hopkin: Keeping these experimental vocalizations to a whisper makes sense if you want to avoid attracting predators to your nest. And South American parrots are not the only ones that do it. Berg has since found that the babies of another type of parrot, one that lives in Mexico and southern Texas, also babble–in work he'll be presenting at the Animal Behavior Society's annual meeting this summer.

[CLIP: Parrot babbling]

Hopkin: Now, as to what all this babbling is about, Berg says it's not likely a form of social communication, because the birds often do it when no one else is listening.

Berg: A better explanation for what they're doing is: they're practicing and possibly adjusting the way they make certain calls as they get a better handle on what they're hearing, presumably by watching and listening to adults as they interact.

Hopkin: So, in addition to producing the standard "feed me" chirps, the babies are tossing out contact calls like the ones adults use to coordinate their movements, alarm calls like the ones that ring out when a parrot spots a snake or a hawk and warbling calls that are produced to defend the nest. But in babbling, all these calls are just sort of jumbled together without any readily identifiable purpose or context. It's like a rehearsal for when the birds really do have nests to defend, behaviors to coordinate and predators to warn about.

Berg: So babbling is a medley, a tossed salad of all these different calls that either are already or will eventually be used in more specific functional contexts.

Hopkin: Berg also found that a babbling parrot's playlist can be boosted by steroids.

Berg: By giving a modest dose of corticosterone to nestlings twice a day for a week, we were able to show that the corticosterone-treated nestlings had a larger repertoire.

Hopkin: That makes sense because stress hormones are key to coordinating a baby bird's growth and maturation. Now, as to whether having a more extensive songbook might be beneficial ...

Berg: It's quite possible. A large repertoire in songbirds has been shown to be advantageous in numerous ways related to reproduction.

Hopkin: To find out if the same is true for parrots, Berg will continue to observe his oldest feathered friends.

Berg: One way we can figure that out is by continuing the long-term ecological monitoring, where we capture essentially every individual in the population and put unique color bands on them.

Hopkin: That way Berg and his colleagues can assess whether the parrots who had the widest vocabulary in their baby days ...

Berg: Are the movers and shakers in parrotlet society today ...

Hopkin: At least as evidenced by their reproductive success, which would be something to crow about.

[CLIP: Adult Green-rumped Parrotlet calls]

Hopkin: *Science, Quickly* is produced by Jeffery DelViscio, Tulika Bose and Kelso Harper. Subscribe wherever you get your podcasts and visit ScientificAmerican.com for updated and in-depth science news.

[CLIP: Show theme music]

For *Scientific American*'s *Science, Quickly*, I'm Karen Hopkin.

About the Author

Karen Hopkin is a freelance science writer in Somerville, Mass. She holds a doctorate in biochemistry and is a contributor to Scientific American's *60-Second Science podcasts.*

Bumblebees Show Off Their Own Puzzle-Solving Culture

By Rachel Nuwer

Culture, at its most basic, refers to socially learned behaviors that are shared among a population. Until the mid-20th century, this ability was thought to be something uniquely human. But bountiful evidence now shows that culture exists in a wide variety of species, from bighorn sheep and vervet monkeys to meerkats and cranes.

Scientists are even discovering that insects can join in their own culture. In a new *PLOS Biology* study, researchers used a gold-standard test that's been applied to species such as chimpanzees and great tits to reveal that bumblebees are capable of cultural transmission of information from one insect to another. Bees that were taught one of two solutions for opening a puzzle box spread that behavioral trait to untrained bees, creating a cultural signature for their colony. "This is an animal with a brain the size of a pinhead, and still they could achieve similar things [as] primates or birds, which is quite remarkable," says lead author Alice Bridges, now a lecturer in biology and animal behavior at Anglia Ruskin University in England, who undertook the study as part of her doctoral research at Queen Mary University of London.

Insects are known to be capable of social learning in the wild: the best-known example is the so-called waggle dance used by honeybees to communicate the location and quality of flowers. The new study builds on prior findings that bumblebees can learn complex behaviors in the lab, too, such as accessing rewards by pulling strings or rolling balls to play a sort of insect version of soccer. These earlier studies showed that bumblebees were capable of social learning but not necessarily of culture, which involves the spread of specific behaviors throughout a population. "We wanted to see whether the bees would be able to maintain a cultural tradition, even when there was an alternative behavior they could do," Bridges says.

She and her colleagues adapted a puzzle box design that was previously used in experiments with primates and birds for their grape-sized study subjects. The bees could either push a red tab clockwise or a blue tab counterclockwise to access a sugar solution contained on a yellow target below. The researchers selected demonstrator bees from four colonies to teach others and trained half of these demonstrators to access the reward under the red tab and half to access it under the blue tab. Then the team put the bees back into their respective colonies, along with access to additional puzzle boxes.

As seen in other species, the researchers observed that the demonstrators proceeded to open the boxes in the way they had been taught and that a subset of those bees' colony mates learned to do the same by watching them. Bumblebees, like people, have different personalities when it comes to work, from highly motivated to lazy, Bridges says, and bees in the former group tended to be the ones that learned the new behavior. Some go-getter individuals took just one day to begin mimicking the demonstrators' behavior and solving the puzzle, whereas others took up to four days. On occasion, a bee would stumble on the solution for opening the other colored tab that the demonstrators in its colony hadn't been taught to push. But the researchers found that this behavior would not stick and that those individuals would always go back to opening the tab that corresponded to their demonstrator bees' behavior–that is, to their colony's culture.

In two control colonies in which bees were trained to recognize the yellow target as signaling a reward but no demonstrators were trained to solve the puzzle, some individual insects did figure out how to open the box on their own. They were not as proficient as bees in experimental colonies that learned from demonstrators, however. Of the control bees that did figure this out, "most did it only a handful of times and then never again," Bridges says. "So having the social learning element was critical to get the bees to actually permanently add box opening to their behavioral repertoire."

This "exciting and groundbreaking study" takes the evidence that culture is widespread in the animal kingdom "to new levels," says Andrew Whiten, a cognitive ethologist at the University of St Andrews in Scotland, who was not involved in the research. "The authors were coy in not venturing to put the word 'culture' in their title, but if it had been me, I would have had it there."

Bridges suspects that other social insects such as honeybees and leaf cutter ants would show a similar capacity for culture. As scientists continue to add species to the list of those that demonstrate such capability, these new findings make a case, she says, that culture probably isn't the "pinnacle of evolution" that humans have previously assumed it to be.

"Maybe culture is not something that is super exclusive or difficult," Bridges says. "Maybe culture really is not all that."

About the Author

Rachel Nuwer is a freelance science journalist and author who regularly contributes to Scientific American, *the* New York Times *and* National Geographic, *among other publications. Follow Nuwer on Twitter @RachelNuwer.*

What Scientists Are Learning about Women's Health from Other Female Animals

By Barbara Natterson-Horowitz

Projected on the massive screen behind me onstage, a herd of giraffes rushes across a sweep of savanna. With the video set to loop, the giraffes gallop endlessly, giving me time to slowly lean across the podium and ask my audience: "Did you spot the pregnant giraffes?" I am delivering a plenary lecture at the 2019 Nobel Conference in Stockholm. The theme of that year's conference was bioinspired medicine–finding solutions in nature to human health problems–and I wanted to call attention to the connections between women and other female animals.

As a cardiologist and evolutionary biologist, I'd been posing this question about the giraffes to medical students in my courses at Harvard University and the University of California, Los Angeles, for years, so I could tell it had landed as planned. I watched the crowd scan the troop of giraffes for evidence of pregnancy–a baby bump, a lagging mother-to-be. I suspected that few, if any, of the assembled scientists and physicians had considered this question when first taking in the scene. That was precisely my point. Given the importance of female health challenges such as pregnancy to the survival of a species–including our own–shouldn't the realities of female life in the wild be more than an afterthought for doctors and biomedical researchers?

Predators pose a daily threat to survival for all prey species, and they don't give pregnant animals a pass. Even in their final, heaviest days of pregnancy, females must evade predators. To do so, they have evolved impressive physiological adaptations.

Consider a giraffe in the last weeks of her nearly 15-month gestation. She has gained hundreds of pounds–her fetus alone weighs up to 150 pounds, but the load is much heavier when you add in the placenta, extra fluid and fat. Yet pregnant giraffes appear to flee as fast as nonpregnant individuals. If they couldn't, predators would quickly target them as easy prey. That would be the end of the line for that mother, her gestating calf and ultimately the species itself.

As I publicly expressed admiration for the giraffe's gestational athleticism, I privately cringed a bit recalling how unathletic–how un-giraffe-like–I had felt as a cardiologist in the final weeks of my own pregnancies. I was the slowest doctor in the herd of physicians responding to code blue alerts for medical emergencies at the hospital. I'd become so breathless and exhausted rushing up just two flights of stairs that once, on my way to a cardiac arrest, another physician pulled me aside to express concern for my cardiac health.

I understood why they were worried. Late in pregnancy, women are at risk of developing life-threatening diseases such as preeclampsia and heart failure. Even a healthy pregnancy places significant stress on the heart. The volume of blood that is circulating expands nearly 50 percent. To deal with this increased workload, the cardiac cells and pumping chambers of the heart must grow and transform in size and shape. Cardiologists use the term "remodeling" to describe these changes, but when things go wrong, the consequences are far more serious than a misplaced sofa throwing off the feng shui. Problems with how the heart cells get remodeled during pregnancy are linked to reduced cardiac function and even heart failure soon after delivery.

Fortunately, my cardiac health turned out to be fine: the source of my shortness of breath and fatigue was the nearly 40 pounds of fetus, fluid and fat I'd packed on by the end of the third trimester. Slowing down in the last weeks of pregnancy is perfectly normal for humans–including female cardiologists who can use the elevator instead of sprinting up the stairs.

But in the wild, pregnant animals must maintain their speed and stamina to save their lives. For pregnant gazelles, zebras, and other prey species, the ability to flee swiftly protects them from predation. The animals giving chase are often also female and sometimes also pregnant. To avoid starvation, pregnant cheetahs, hyenas, and other predators need to pursue, overtake and capture fleeing prey. A species in which pregnant females can't evade predators or capture food is doomed to extinction.

I've become fascinated by how females of other species have adapted to these challenges. I now study their biology to find possible solutions to women's health issues ranging from heart failure to breast cancer. This field of research is still emerging, but already it is generating insights that could lead to lifesaving treatments for some of the most prevalent human health concerns.

Missing Females

A multispecies approach never would have occurred to me as a young professor of medicine in the 1990s. For the first 10 years of my career, I treated many forms of heart disease but always in one species—my own. Then, in 2005, I became a cardiovascular consultant to the Los Angeles Zoo. (Zoo veterinarians occasionally invite physicians to weigh in on diagnoses or assist in medical procedures, especially when the conditions are far more common in humans than in nonhuman animals.) My patient roster expanded to include great apes, bears, lions and raptors, among others. Animal health hadn't been included in my formal medical education, so the learning curve was steep. The zoo's veterinarians and the animals themselves became my teachers as I learned that most of the diseases affecting my human patients—cardiovascular disease and cancer among them—can also develop in other species.

Over the years I grew increasingly convinced that species-spanning medicine could reveal insights for my patients that traditional human-centered approaches had not. Partnering with naturalist and science writer Kathryn Bowers, I made it my mission to raise awareness about the relevance of animal health to that of humans, launching educational and professional programs, leading research teams and writing books meant to bring the fields of human medicine, veterinary science and evolutionary biology closer together. The reception to these efforts was mostly positive, but I received a surprising degree of pushback from my own field of medicine. Some physicians seemed resistant to connecting the health of their patients and other species (beyond the transmission of certain diseases from animals to humans). Maybe they took issue with the humbling notion that veterinarians could teach them about their human patients.

Human exceptionalism, I could see, was deeply entrenched in our medical traditions. People haven't always been comfortable accepting that humans are animals. It has been my experience that human health professionals reflexively assume diseases they treat in their patients are unique to our species. (They aren't.)

Anthropocentrism wasn't the only distorted worldview slowing down the progress of medical science. Biomedical research has also long focused on males to the exclusion of females. Until 1993 the National Institutes of Health didn't require the clinical trials they funded to include women—not even studies of diseases known to be more common in women than men. Mind-bogglingly, it wasn't until 2015 that the NIH mandated the inclusion of female laboratory animals in preclinical safety studies of new medical treatments. There were rationales for not including females in biomedical studies: researchers wanted to avoid the potentially confounding effects of the estrus cycle, which can introduce variations in physiology and behavior not seen in males, and they wanted to protect potentially pregnant women, and fetuses, from thalidomide-type tragedies. Yet however well intended these investigators might have been, excluding females from studies proved enormously damaging to women's health.

To address the gaps in our medical knowledge, I began to focus my species-spanning research on major challenges in women's health. The health connections between human and nonhuman females weren't hard to find. I've published studies on ovarian cancer in flamingoes, pythons, fish and humans; hosted symposia on menstrual difficulties in great apes (including us), bats and tree shrews; and collaborated with dairy veterinarians who have deep knowledge of lactation in cows to help women with common breastfeeding problems. These experiences have transformed my understanding not only of my patients but of myself as a female. I've learned when it comes to certain aspects of my health, I may have more in common with other female animals than with my husband, brother, son or the other men in my life. I call the shared bond that links human and animal females the "sisterhood of species."

Lions Get Breast Cancer

I first glimpsed the connection among female animals while examining Cookie, a geriatric lioness, at the Los Angeles Zoo in 2012. After an operation, a lion's heartbeats sound pretty much the same as those of a human who has been through the same procedure. So the clarity of the lub-dup, lub-dups streaming through the stethoscope I had pressed against Cookie's huge, furry chest reassured me. It meant the dangerous buildup of fluid we had removed from the sac around her heart hadn't returned. The pathology report on the cause of the fluid buildup wasn't back yet, however, and the veterinarians were worried that it might be metastatic breast cancer. There were two reasons for this specific concern. First, breast cancer is one of the tumors that sometimes spreads to the heart, creating inflammation and the kind of fluid buildup that had threatened Cookie's life. The other reason, the veterinarians explained, was that compared with other zoo animals, lions, jaguars, cheetahs, and other cats are especially susceptible to breast cancer.

This connection between human and animal health was personal. Breast cancer is a health concern for many women, and I was no exception. Around the time I first learned about breast cancer in cats, I was being tested for the *BRCA* mutations following several borderline breast biopsies. The *BRCA* mutations are well known to put women at increased risk of breast and ovarian cancers. To my amazement, I learned that *BRCA* mutations also put some dog breeds, including Cavalier King Charles Spaniels, and certain other carnivores at increased risk. In all my years caring for women, as well as some men, with breast cancer, I hadn't considered what other species might also be vulnerable. In fact, as I showed in a recently published study, breast cancer is a threat across nearly all mammalian lineages, with cases found in creatures ranging from kangaroos and koalas to wombats and water buffalos.

Ultimately Cookie and I were both lucky. Cookie's fluid buildup turned out not to be cancerous, and she went on to live another four years. I tested negative for *BRCA* mutations, and my biopsies did not show breast cancer. Other creatures have not fared so well. But we have learned from their misfortune.

Whales don't usually develop breast cancer–leukemia and lymphoma are the more common cetacean cancers. So when researchers found that 27 percent of the dead belugas they examined in and around the St. Lawrence Estuary near Montreal between 1983 and 1999 had died of breast cancer, their discovery raised questions.

Looking for causes, investigators quickly locked on to several aluminum smelting factories near the estuary that were spewing chemicals known to induce breast and other cancers. The scientists determined that these chemicals probably contributed to the disease not only in the whales but also in humans who lived in the area and exhibited abnormally high cancer rates themselves. Recognizing the breast cancer "epidemic" in mammals living in this urban waterway helped to alert authorities to an environmental hazard that was also threatening human health.

Like the canaries that British coal miners brought with them to detect the presence of hazardous gases in the mines, the female belugas signaled that contaminants were putting humans at risk. Today rising levels of pollution are implicated in many challenges to women's health, including cancers, infertility, premature labor, and other disorders of the female reproductive system. Nonhuman female animals living around human communities are increasingly exposed to the same pollutants as women and girls, as the lines that once separated human and animal environments become blurred. Simply stated, the health of female animals is too important to our own to ignore. All female animals–including women and girls–are now canaries for one another. Our coal mine is the planet we share.

Giraffes Avoid Heart Failure

The connection among female animals extends beyond shared vulnerabilities. Some members of the sisterhood have evolved unique physiologies to defend against disease. The remarkable biology of some female animals prevents them from falling prey to diseases that claim millions of women's lives every year. To understand how and why such unique biology evolved and how it could save women's lives, consider my new take on the evolution of the giraffe and its improbably long neck.

Rewind to 11.5 million years ago, when the common ancestors of the modern giraffe and its closest evolutionary cousin, the okapi, roamed what is now Africa. Some of those ancestors wandered toward the rain forest and present-day Congo and eventually gave rise to the okapi. Other ancestors migrated toward the savanna and gave rise to the giraffe. Those differing environments helped to produce very different necks in these two lineages. In the rain forest, okapis were able to survive and thrive with perfectly fine albeit unspectacular necks. The ancestral giraffe's environments, in contrast, lacked dense foliage for hiding. A longer neck would have permitted horizon scanning and earlier detection of predators in an open grassland setting. At the same time, getting taller may have given proto-giraffes better access to sometimes limited foliage.

A downside to having a neck that places the head more than 2.5 meters away from the heart is that this all-important muscle must work harder with each beat to push blood vertically to the brain. This greater pressure on the heart is what we call blood pressure, and in a healthy giraffe, systolic blood pressure (the pressure when the heart is contracting) can exceed 250 millimeters of mercury (mm HG), compared with 120 mm HG or less for a healthy human.

Generally the problem with high blood pressure is that when a muscle like the heart works harder, it thickens. In humans and mice, when hearts thicken this way, they develop scarlike tissue, known as fibrosis. Hearts become stiffer than normal, limiting how far or fast an individual can move. To make matters worse, high blood pressure, also known as hypertension, can damage and weaken the heart over time. In humans, it results in a condition called HFpEF (for heart failure with preserved ejection fraction, pronounced "hef-pef"), which leads to exhaustion, shortness of breath and, potentially, death. HFpEF is the leading cause of heart failure in women, responsible for tens of thousands of deaths a year in the U.S. alone.

On the wide-open savanna, such stiffening could slow an animal down, threatening its survival. Hello, lion; goodbye, giraffe. But giraffes have found a workaround. Despite having blood pressure well above the levels that would earn human patients a stern talking-to from their doctor, both male and female giraffes can gallop at lion-evading speeds, and they appear to be resistant to HFpEF. Recent genomic research hints at how giraffes are able to flout the rules that constrain shorter animals.

In the past few years Chang Liu and his colleagues at Northwestern Polytechnical University in Xi'an, China, have published high-quality genomes of giraffe, okapi and other related animals. They discovered several mutations in the giraffe genome that are not found in the other species. One of the affected genes, *FGFRL1*, is linked to high blood pressure, which can lead to heart failure in humans. Could giraffe *FGFRL1* explain the longer-necked species' resistance to diseases associated with increased blood pressure?

To find out, the scientists tested two groups of mice to see how each would respond to an infusion of a hormone that causes blood pressure to rise. The first group of mice were "wild type," meaning that all their genes, including *FGFRL1*, were mouse genes. The second group had their *FGFRL1* sequence edited with CRISPR technology to mimic a giraffe's. After 28 days of exposure to the blood pressure–raising chemical, the wild-type mice had high levels of fibrosis, a common finding in humans with heart failure caused by hypertension. The hearts of the mice with giraffe *FGFRL1*, in contrast, appeared nearly fibrosis-free. This finding suggests that giraffes may have evolved a unique ability to suppress some of the damaging fibrosis of the heart.

Giraffes also appear to have evolved mechanisms to shield against other damaging effects of hypertension, including when it occurs during pregnancy. In humans, gestational hypertension–having high blood pressure during pregnancy–is pathological. A leading cause of fetal and maternal mortality, it occurs in up to 25 percent of pregnancies worldwide and is on the rise. For pregnant giraffes, in contrast, higher blood pressure is normal, and neither mothers nor babies seem to suffer any consequences from it. I've brought together a team of veterinarians, pathologists and OB-GYNs to uncover the unique biology that protects giraffes from gestational hypertension. We are comparing the placentas of giraffe and okapi, looking for clues to the evolution of this trait.

It will take time to figure out how to use the knowledge we obtain from such studies of female animals to prevent diseases in women. The principle is clear, though: many of the greatest threats to women's health may have already been neutralized in other species. The fixes are out there, in the bodies of the animals with whom we share our planet. As ingenious as we humans may be, the natural world is even smarter. Evolution's ability to solve problems is, according to Charles Darwin himself, "immeasurably superior to man's feeble efforts."

Rachel Carson is one of my scientific heroes, not only because she launched the modern environmental movement but also because she recognized the powerful connections across species. "In nature," she wrote, "nothing stands alone." I would add that no female stands alone. Members of the sisterhood of species are linked by common ancestry and the shared challenges and joys of being a female animal.

About the Author

Barbara Natterson-Horowitz is a cardiologist and evolutionary biologist on the faculties of Harvard University and the University of California, Los Angeles. She studies evolved adaptations in female animals that could improve women's health.

Humans Are Not the Only Creatures Who Mourn

By Barbara J. King

Animal behaviorists have traditionally shied away from attributing human emotions, such as grief, to responses by animals.

But a growing body of evidence indicates that species ranging from dogs to dolphins mourn the passing of relatives and close companions.

These observations suggest that although the ways in which we mourn may be uniquely human, our capacity for grief has deep evolutionary roots.

On a research vessel in the waters off Greece's Amvrakikos Gulf, Joan Gonzalvo watched a female bottlenose dolphin in obvious distress. Over and over again, the dolphin pushed a newborn calf, almost certainly her own, away from the observers' boat and against the current with her snout and pectoral fins. It was as if she wanted to nudge her baby into motion—but to no avail. The baby was dead. Floating under direct sunlight on a hot day, its body quickly began to decay; occasionally the mother removed pieces of dead skin and loose tissue from the corpse.

When the female dolphin continued to behave in this way into a second day, Gonzalvo and his colleagues on the boat grew concerned: in addition to fussing with the calf, she was not eating normally, behavior that could be risky for her health, given dolphins' high metabolism. Three other dolphins from the Amvrakikos population of about 150 approached the pair, but none disrupted the mother's behavior or followed suit.

As he watched the event unfold in 2007, Gonzalvo, a marine biologist at the Tethys Research Institute in Milan, Italy, decided he would not collect the infant's body to perform a necropsy, as he would usually have done for research purposes. "What prompted me not to interfere was respect," he told me in 2013. "We were privileged to be able to witness such clear evidence of the mother-calf bond in bottlenose dolphins, a species that I have been studying for over a decade. I was more interested in observing that natural behavior than interrupting it by abruptly interfering and disturbing a mother who was already in obvious distress. I would define what I saw as mourning."

Was the dolphin mother truly grieving for her dead calf? A decade ago I would have said no. As a biological anthropologist who studies animal cognition and emotion, I would have recognized the poignancy of the mother's behavior but resisted interpreting it as mourning. Like most animal behaviorists, I was trained to describe such reactions in neutral terms such as "altered behavior in response to another's death." After all, the mother might have become agitated only because the strange, inert status of her calf puzzled her. Tradition dictates that it is soft-hearted and unscientific to project human emotions such as grief onto other animals.

I think Gonzalvo was correct in his judgment that the mother dolphin was mourning. Indeed, in 2016 Melissa Reggente of the University of Milano-Bicocca and her colleagues published a stunning report that described 14 responses to death in seven species of whales and dolphins, including clear evidence of mourning. In the Red Sea near Egypt, for instance, an adult bottlenose dolphin repeatedly touched the body of a juvenile dolphin and pushed it in the water. Taken together, the new examples provide a clear window on grieving in cetacean cultures. In the past few years, in fact, a critical mass of new observations of animal responses to death has bubbled to the surface, leading me to a startling conclusion: cetaceans, great apes, elephants, and a host of other species ranging from farm animals to domestic pets may, depending on circumstances and their own individual personalities, grieve when a relative or close friend dies. That such a broad range of species–including some quite distantly related to humans–lament the passing of loved ones hints that the roots of our own capacity for grief run very deep indeed.

Defining Grief

Since Charles Darwin's day, two centuries ago, scientists have debated hotly whether some animals display emotions beyond those associated with parental care or other aspects of survival and reproduction. Darwin thought that, given the evolutionary connection between humans and other animals, many emotions must be similar across species. He granted to monkeys, for instance, grief and jealousy, as well as pleasure and vexation. But the attribution of emotions such as these to animals fell increasingly out of mainstream scientific favor. By the early 20th century the behaviorist paradigm held sway, with its insistence that only observable behavior of animals, not their interior lives, could be studied with rigor. Gradually the scientific embrace of animal emotion has revived, thanks originally in part to anecdotes from long-term field studies on large-brained mammals. From Tanzania, Jane Goodall recounted in heart-wrenching detail young chimpanzee Flint's decline and death from grief only weeks after the death of his mother, Flo. From Kenya, Cynthia Moss reported that elephants attend to dying comrades and stroke the bones of deceased relatives. Field biologists and anthropologists began to ask questions about whether, and how, animals mourn.

To study and understand grief among animals, scientists need a definition that distinguishes it from other emotions. Whereas "animal response to death" embraces any behavior by an individual following the death of a companion animal, researchers may strongly suspect grief only when certain conditions are met. First, two (or more) animals choose to spend time together beyond survival-oriented behaviors such as foraging or mating. Second, when one animal dies, the survivor alters his or her normal behavioral routine–perhaps reducing the amount of time devoted to eating or sleeping, adopting a body posture or facial expression indicative of depression or agitation, or generally failing to thrive. For his part, Darwin conflated grief with sadness. But the two differ, primarily in intensity: the grieving animal is more acutely distressed, possibly for a more prolonged period.

This two-part definition is imperfect. For one thing, scientists lack a metric for evaluating exactly what counts as "more acutely distressed." Should the criteria for grief differ according to species? Might grief in other animals assume forms that are difficult for humans to recognize as mourning? Data are not yet available on these questions. Further, mothers or other caretakers that constantly provide food or protection to infants that subsequently die cannot be said to have met the first criterion (going beyond survival-oriented behaviors), yet they remain among the strongest candidates for suffering survivor's grief.

Future studies of animal mourning will help refine this definition. For now, it furthers our critical assessment of responses made by animals when others around them die. For instance, baboon and chimpanzee mothers in wild African populations sometimes carry the corpse of their dead babies for days, weeks or even months—a behavior that on the surface of things might look like grief. But they may not exhibit any significant outward indicator of agitation or distress. When the animals carry on with their routine behaviors, such as mating, their behavior does not meet the criteria for mourning.

A Menagerie of Mourners

A wide range of species do exhibit behaviors that fit the two-part definition of grief, however, elephants among them. A particularly compelling example of elephant mourning comes from Iain Douglas-Hamilton of Save the Elephants and his team at Kenya's Samburu National Reserve, who in 2003 tracked elephants' responses to the dying matriarch called Eleanor. When Eleanor collapsed, a matriarch named Grace from another elephant family immediately came to her aid, using her tusks to support Eleanor back onto her feet. When Eleanor fell again, Grace stayed with her, pushing on her body, for at least an hour, even though her own family moved on. Then Eleanor died. During the course of the week that followed, females from five elephant families, including Eleanor's own, showed keen interest in the body. Some individuals appeared upset, pulling at and nudging the body with trunk and feet or rocking back and forth while standing over it. Based on the females' reactions (at no point during this period did a bull elephant visit the carcass), Douglas-Hamilton concluded that elephants show a so-called generalized response to dying and death—grieving not only for the loss of close kin but for individuals in other families.

Wild cetaceans also seem to exhibit a generalized grief response. In 2001 in the Canary Islands, Fabian Ritter of Mammal Encounters Education Research observed a rough-toothed dolphin mother pushing and retrieving her dead calf's body in much the same way that the Amvrakikos dolphin mother later did with her baby's corpse. She was not alone: two adult escorts swam synchronously with her at certain periods, and at other times a group of at least 15 dolphins altered their pace of travel to include the mother and dead baby. The mother's persistence was remarkable, and when on the fifth day it began to wane, the escorts joined in and supported the infant on their own backs.

Giraffes, too, appear to grieve. In 2010 at the Soysambu Conservancy in Kenya, a Rothschild's giraffe gave birth to a baby with a deformed foot. The baby walked less and remained more stationary than most calves. During the youngster's four weeks of life, wildlife biologist Zoe Muller, then at the Rothschild's Giraffe Project, based in Kenya, never saw the mother more than 20 meters away. Although individuals in a giraffe herd often synchronize their activities, foraging together, for example, the mother deviated from this pattern, preferring to stay close to the baby. Like the dolphin mother in the Amvrakikos Gulf, she may have risked her own health in doing so—though in this case for a living offspring.

One day Muller discovered the herd engaged in highly atypical behavior. Seventeen females, including the calf's mother, were vigilant and restless as they stared into a patch of bush. The calf had died in that spot about an hour before. All 17 females showed keen interest in the body that morning, approaching and then retreating from it. By the afternoon 23 females and four juveniles were involved, and some nudged the carcass with their muzzles. That evening 15 adult females clustered closely around the body—more closely than they had been during the day.

Throughout the following day numerous adult giraffes attended the infant's body. Some adult males approached for the first time, although they showed no interest in the carcass, instead focusing on foraging or inspecting the reproductive status of the females. On day three Muller spotted the mother giraffe alone under a tree about 50 meters from where the calf had died. The body itself, however, was no longer in its resting spot. Following a search, Muller located it, half-devoured, in the spot under the tree where the mother had been earlier. By the next day the body was gone, taken by hyenas.

Giraffes are highly social animals. After caching a newborn out of sight for about the first four weeks of life, the mothers sometimes engage in a crèche system in which one looks after the infants while the others forage. Muller does not use the word "grief" or "mourning" in describing the incident she witnessed. Yet this case is especially instructive. Not only the mother's behavior but also that of many of the females in her herd changed significantly in the wake of the infant's death. Although it is impossible to rule out an alternative explanation, the fact that the females had mounted a protective response against predators taking the baby makes it overwhelmingly likely that grief was involved.

Detailed observations of wild populations of animals, such as the ones Muller reported, are still relatively rare, for several reasons. Scientists may not be at the right place at the right time to observe postdeath responses by survivors. And even when they are present, no remarkable grief behaviors may ensue. Especially at this early stage of research into animal grief, observations from sanctuaries, zoos and even our own homes may supply needed clues.

I cannot imagine describing the behavior of Willa the Siamese cat without invoking the word "grief." For 14 years Willa lived with her sister, Carson, at the home of Karen and Ron Flowe in Virginia. The feline siblings groomed each other, lazed together in favorite parts of the house and slept with their bodies entwined. If Carson was taken from the house to visit the vet, Willa acted mildly agitated until she reunited with her sister. In 2011 Carson's chronic medical issues worsened, and the Flowes took her again to the vet, where she died in her sleep. At first, Willa acted as she did when her sister was away for a brief period. Within two or three days, though, she began to utter an unearthly sound, a sort of wail, and to search the spots she and Carson had favored together. Even when this startling behavior faded, Willa remained lethargic for months.

The loyalty of dogs to their human companions is legendary, and so may be their grief. Visitors to Tokyo sometimes make a pilgrimage to the Shibuya train station to view a statue of an Akita named Hachikō, who waited daily for more than a decade for his human friend Hidesaburō Ueno to step off a train. Ueno, a professor who had died suddenly at his university office, would never again walk home with Hachi (as the dog was nicknamed), but the dog maintained a hopeful vigil until his own death. This profound loyalty is also found in dog-dog friendships. The ASPCA's Companion Animal Mourning Project reports that two thirds of dogs show behavioral changes indicative of emotional upset, lasting up to six months, when another dog in the same household dies.

These statistics reflect my family's experience. My cousin Connie Hoskinson's silky terrier, Sydney, grieved twice over: for Connie's husband, George, when he died and later when her second dog, Angel, adopted when Sydney was 13 years old, suffered a sudden fatal heart attack. Without his canine friend–with whom he had slept and for whom he searched the house for weeks–Sydney ate less, lost weight and before long died in his sleep.

The Sorrow Continuum

It is logical to think that long-lived species whose members partner most closely with others in tight-knit pairs, family groups or communities may more readily mourn the deaths of loved ones than other species do. But researchers do not yet know enough about animal grief to make such a claim. We need to test this hypothesis by systematically comparing responses to death in a variety of animal social systems, from gregarious ones to those in which animals come together only seasonally for food or mating.

Still, species-level differences in grieving will not be the whole story, because variation in the immediate social contexts and personalities of individual survivors will complicate matters. For instance, whereas the practice of allowing a survivor to view the body sometimes seems to prevent or reduce a period of distressed searching and vocalizing by the surviving animal, other times it seems not to help at all–attesting to the degree of individual variation in death responses within species. Likewise, evidence for grief in wild monkeys that live in cohesive social units is surprisingly limited so far, whereas in more solitary species such as domestic cats, bonds may develop between two or more kin or friends such that grief responses rival those of much more social animals. Some instances of grief in wild monkeys are coming to light, though, including in South American marmosets, Sichuan snub-nosed monkeys, and Barbary macaques. In my book *How Animals Grieve*, I recount examples from rabbits, horses and birds, as well as the other animals discussed here. In each species I find a grief continuum, with some individuals seeming indifferent to a companion's death and other individuals appearing distraught over such a loss.

Cognitive differences also play a role in animal grief. Just as there are different levels of empathy expressed by different species and even across individuals within a species, there must be varying levels of comprehension when animals grieve. Do some animals grasp death's finality or even have a mental concept of death? We simply don't know. No evidence suggests that any nonhuman animal anticipates death in the way we humans do, a capacity that underlies so much of our compelling literature, music, art and theater–and that causes our species a great deal of emotional suffering.

Indeed, the capacity to mourn may become quite costly for any animal in both physical and emotional terms, especially in the wild where alert high-energy behavior is needed for foraging, predator avoidance and mating. Why then did grief evolve in the first place?

Perhaps the social withdrawal that often accompanies an animal's grief, if not taken too far, allows time for rest and thus an emotional recovery that leads to greater success in forging a new close bond. Or, as John Archer writes in *The Nature of Grief*, it may be that "the costs involved in grief can be viewed as a trade-off with the overall benefits conferred by separation responses" seen when two individuals are keenly attached but forced apart from each other.

Under such circumstances, the missing partners may search for each other and thereby reunite and live to see another day. What is adaptive, then, may not be grief itself but instead the strong positive emotions experienced before grief comes into the picture, shared between two or more living animals whose level of cooperation in nurturing or resource-acquisition tasks is enhanced by these feelings.

The Price of Love

From this perspective, we may link grief with love, full stop. That is to say, grief results from love lost. Exploring emotions in a variety of species, ecologist and animal behaviorist Marc Bekoff of the University of Colorado Boulder embraces the idea that many animals feel "love" as well as "grief," even as he acknowledges that those concepts are hard to define precisely. We humans, he notes, do not fully understand love, but we do not deny its existence—or its power to shape our emotional responses.

In his book *Animals Matter*, Bekoff tells the story of a coyote called Mom whom he observed for several years during behavioral studies in Wyoming's Grand Teton National Park. At one point Mom began to make short journeys on her own away from her pack. Her offspring would rejoice when she returned: they licked Mom and rolled over exuberantly at her feet. Then Mom left for good. Some of the coyotes in her pack paced; others searched for her, setting off in the direction Mom had departed. "Her family missed her," Bekoff wrote. Discussing this with me, Bekoff attributed the family's response to its love for Mom. Generally, the potential for love is strong in species such as coyotes, wolves and many birds, he said, because male and female partners defend territories, raise their young together, and miss each other when they are apart.

Love in the animal world often entwines with grief in an acute mutuality. Perhaps even more than the degree of social cohesion within a species, it is love between individuals that predicts when grief will be expressed. Can there be any real doubt that Willa, a representative of a species (the domestic cat) not known for its social nature, loved her sister, Carson, or that as the sole surviving sister, she suffered grief in the wake of her loss?

In our own species, grief increasingly became expressed through rituals rich in symbolism. By around 100,000 years ago, our *Homo sapiens* ancestors decorated dead bodies in red ocher, a behavior interpreted by archaeologists to be a kind of symbolic (rather than functional) ornamentation. At a site in Russia called Sunghir, two children younger than 13 years, a boy and a girl, were buried 24,000 years ago, together with elaborate grave goods ranging from mammoth tusks to animals carved from ivory. Most astonishing were the thousands of ivory beads found in the pair's grave, probably sewn onto the clothing (long since disintegrated) in which the children were buried. A good portion of this ancient human community at Sunghir must have come together in preparing this funeral ritual—each bead alone took an hour or more to manufacture. Although it is risky to project modern emotions onto past populations, the examples of animal grief reviewed here strengthen an emotion-based interpretation of the archaeological evidence: our ancestors of many thousands of years ago mourned their lost children.

In our modern world, grief is no longer inevitably confined to kin, close social partners or immediate members of one's own community. Public commemoratives at the Peace Memorial Park in Hiroshima; the genocide memorial center in Kigali, Rwanda; the Foundation Memorial to the Murdered Jews of Europe in Berlin; the site of the Twin Towers in Manhattan; or the Sandy Hook Elementary School in Newtown, Conn., all convey visibly the power of agonized global mourning. Our human capacity for sorrow at the deaths of strangers is built on an evolutionary substrate. Our own ways of mourning may be unique, but the human capacity to grieve deeply is something we share with other animals.

Referenced

How Animals Grieve. Barbara J. King. University of Chicago Press, 2013.

Mind Your Moods, Cat Owners. Barbara J. King. NPR. January 22, 2015. www.npr.org/sections/13.7/2015/01/22/379008858/mind-your-moods-cat-owners

About the Author

Barbara J. King is emerita professor of anthropology at the College of William and Mary. Her studies of monkeys and apes have led her to examine emotion and intelligence in a wide range of animal species.

The Social Genius of Animals

By Katherine Harmon Courage

At the Thai Elephant Conservation Center, tucked away in the trees near Chiang Mai, a pair of Asian elephants gazes at two bowls of corn on the other side of a net. The corn is attached to a sliding platform, through which researchers have threaded a rope. The rope's ends lie on the elephants' side of the net. If only one elephant pulls an end, the rope slides out of the contraption. To bring the food within trunk's reach, the elephants have to do something only humans and other primates were thought to do: they must cooperate. Working in synchrony, each elephant grabs its end of the rope in its trunk and pulls, drawing the platform and the treats within reach.

Six pairs of these large animals succeeded in solving this double rope puzzle. A lone elephant would wait as long as 45 seconds for a partner to arrive, showing it knew it needed a buddy to get the job done. Psychologist Joshua M. Plotnik, now at Mahidol University in Thailand, and his colleagues, who documented these findings in 2011, also noticed that the elephant duos used different strategies for obtaining the food, suggesting that the animals had developed a deep understanding of social cooperation.

In the past century animals have repeatedly surpassed our expectations for their intellect. Koko the gorilla learned sign language; Alex the African grey parrot boasted a spoken vocabulary of more than 150 words, and even invertebrates such as the octopus have been seen using tools in the wild. Until the mid-20th century, we assumed only humans could use tools and learn meaningful pieces of a language. Sophisticated social skills such as cooperation were presumed to be way beyond animals' cognitive reach. Only recently have researchers begun to realize the extent of animals' social intelligence, their ability to understand and learn from others.

The latest findings suggest that some animals are not only capable of creating social connections but that many use them for survival. As with people, some animals have social needs. They notice when another member of their species is distracted, and they are able to figure out an effective way to get its attention. Animals may teach one another important skills, such as how to use tools. Some critters can even deceive. Intuiting what others know enables them to trick one of their kind for personal gain. The animals described below have some of the most impressive social abilities in the animal kingdom. Studying how they interact is giving scientists new insight into what they know–and a glimpse at what might be the upper limits of animal intelligence.

Making Connections

In our own lives, social intelligence might seem to play second fiddle to, say, the ability to do calculus or write a book. Individuals held up as modern geniuses, however, such as Stephen Hawking or the late Steve Jobs, have not only exceptional analytic skills but also an extraordinary ability to relate their ideas to a broader audience. These social skills are fundamental to our complex society. Without the ability to form a cohesive group, communicate ideas or collectively solve problems, humans could not have built the great pyramids or the first supercomputers.

Social skills require, first, a basic affinity for others. "You can't evolve all of the computational systems for social intelligence until you have that tendency that you want to be close to others," says biologist Louis Lefebvre of McGill University. Not all animals have this. Octopuses, for example, seek one another out only to procreate.

In contrast, zebra finches form monogamous pairs and congregate in groups. In a 2009 study the late biologist James Goodson of Indiana University Bloomington and his colleagues drilled down to the chemical essence of this social instinct: mesotocin, the bird equivalent of the human hormone oxytocin, which is implicated in bonding. When the researchers blocked mesotocin in the brains of the birds, they shrugged off their fellow finches. Females spent less than one third as much time near their same-sex cage mates after the treatment than before it. Extra doses of mesotocin had the opposite effect: the birds became even more social than they typically are.

Goodson and his colleagues also examined the biology of this hormone in other species of finches with different degrees of social affinity. The team found fewer receptors for mesotocin in key spots of the brain in species that were more territorial (less social) than the zebra finch–and more receptors in species that traveled in flocks. Scientists have identified similar patterns of brain chemicals and social preferences in many mammals. In prairie voles, for example, how mates bond with one another and their children is also associated with different levels of oxytocin and vasopressin, another hormone linked to social behavior and bonding.

Know Yourself—and Others

Good social graces also require a sense of the self as distinct from others. This primitive level of self-awareness allows animals to go beyond acting out a programmed set of behaviors for mating or for defense.

Researchers test self-awareness by placing a visible mark on an animal and showing it a mirror. An animal passes the test if it appears to recognize that the dot is on its body as opposed to on the "other" animal in the mirror. It can signal this awareness, for example, by reaching toward the spot on its own body. Great apes, elephants, bottlenose dolphins, orcas and magpies have all passed, suggesting they have more social prowess than animals that simply flock or school together.

The second requirement for social networking is an understanding that others have different mental states–knowledge, desires, beliefs or intents–and that at any given moment another individual might be in a mental state that is different from the one you are in. This ability is called theory of mind. Psychologists often test its development in young children by having a child and an adult observe a researcher place an object in a specific location, say, a ball in a cup. The adult leaves the room, and the ball is moved. After the adult returns, a child with a mature theory of mind understands that the adult does not know the ball has moved, and he or she does not expect the adult to look for it in the new location. Similar theory-of-mind tests are difficult to conduct in animals because it is hard to communicate with them directly.

Nevertheless, we can glean hints from their physiology, such as the presence of so-called mirror neurons. These cells, which have been found in macaques and birds, swing into action when an animal sees another individual doing something. In humans, they are thought to exist in various brain regions, including the supplementary motor area, a small patch of tissue at the top of the head that helps to control movement. Because these cells' activity patterns mirror the behavior of others and to some extent reflect the movement's goal, they may enable animals to understand others' actions, possibly even their intentions. They might underlie the ability to learn motor actions by observation and also buttress a theory of mind.

Mirror neurons have not yet been discovered in dogs, but more than a decade ago psychologist Alexandra Horowitz of Barnard College collected behavioral data suggesting that dogs may have some version of a theory of mind. Over the course of 21 months Horowitz videotaped a random sampling of dogs at play in a San Diego dog park. When she analyzed the footage, she noticed some interesting behaviors that suggested dogs could be aware of another dog's perspective.

The dogs at the park varied how they signaled a desire to play depending on the other dog's position. If a prospective playmate was already facing it, the dog gave a visual signal—opening its mouth wide or bowing down into that familiar bottom-up play pose. But if the other dog was turned away and otherwise engaged, it might instead give its friend a little nip. "They seem to be noticing whether a dog they wanted to play with was ready to receive their play signal," Horowitz says. "They seem to be reacting to distinct cognitive states" of others. In a paper published in 2011 Horowitz posited that dogs possess at least a rudimentary theory of mind.

Consistent with the mirror neuron findings, many monkeys and some birds have demonstrated in their behavior at least some capacity for theory of mind. Monkeys, birds and dogs are exceptions, however. In most animal species, scientists have failed to see even a glimmer of evidence for this advanced social capacity.

Schools and Students

Among its many benefits, social intelligence confers the ability to distribute information. Scientists have long observed that herd animals can give warning signals if a predator is approaching; for example, a white-tailed deer may flash its conspicuous white tail, pointing it upward before bolting. Now, however, researchers are noticing that some animals that live in groups can teach one another rules of social engagement or how to create a tool.

Zebra fish, for example, seem to transmit subtle social cues to other members of their school. These small creatures can make an easy meal for larger fish in the wild, so they tend to be cautious of new objects. If they are reared in captivity, however, they show little fear of a moving object, swimming within close proximity of it.

A team of researchers led by biologist Sarah Zala of the Konrad Lorenz Institute of Ethology at the University of Veterinary Medicine in Vienna wanted to know if these fish would change their behavior if surrounded by others that acted differently. The group introduced some shy, wild zebra fish to more brazen domestic ones. The wild ones learned from their new acquaintances to be bolder when a moving object was introduced near the tank. Instead of swimming away from it, they joined their fearless new companions and ventured toward it. When the emboldened wild fish were separated from their new friends, they still bravely swam near novel things. The results, published in June 2012, indicate these fish had not just mindlessly followed the other fish but had learned a new behavior.

Dolphins can convey more intricate behaviors to others. One group of bottlenose dolphins in Australia uses a technique called sponging to find food. Biologist Janet Mann of Georgetown University and her colleagues saw the females sporting a sea sponge on their snout, a technique that seemed to allow the dolphins to push along the seafloor to look for fish and other food without hurting their nose. The scientists sampled DNA from these sponging mammals and compared it with DNA from area dolphins that did not use this tactic. They discovered that the sponging dolphins, but not the others, had a common maternal line. In a 2008 report of the findings, the researchers suggested that mothers in this particular group managed to teach this complex act to their daughters.

Mother chimpanzees also use social learning to hand down wisdom to daughters. Young female chimps at the Gombe National Park in Tanzania are much handier at making tools to retrieve termites than are males. Adult females there use long twigs or stems to fish out termites, whereas males are more likely to kill larger game for their meals. In a study published in 2004, Elizabeth Lonsdorf of the Lincoln Park Zoo in Chicago and her colleagues found that the younger females spent more time with their mothers learning this skill than did the males, who were more likely to be playing—perfecting their social skills for later battles for dominance, mates and food.

Deception

Social learning can be a big advantage for a group or species, which can transmit the most successful strategies to others. Some animals, however, use their social awareness for personal gain by hiding information from potential competitors. A group led by psychologist Federica Amici, then at Liverpool John Moores University in England, trained low-ranking capuchin monkeys, long-tailed macaques and spider monkeys to unlock a box to obtain a food reward. When alone, these monkeys unlocked the box and indulged. But when a dominant monkey was around, the trained animals—especially the macaques—chose to forgo the hidden treat, ignoring the box so as not to reveal how to open it.

Some species of birds engage in a similar type of deceit, bolstering the argument that these animals might also possess some theory of mind. Whether in a laboratory or the wild, Western scrub jays are more likely to move their food stashes when other birds are around. In fact, lab experiments have shown that scrub jays will either move or pretend to move a newly caught worm if they think another bird might have seen them bury it. This behavior suggests some awareness of what the other individual knows and how they can protect their own interests. "Nonprimate species are thinking about these things in some very clever ways," says psychologist Laurie Santos of Yale University.

At least one of the cooperating Asian elephants in Plotnik's experiments also learned how to game the social system. The elephant, named Neua Un, figured out that instead of pulling on her side of the rope, she could stand on her end so it would not slip. Then she let the other elephant do all of the hard work. She also moved her trunk every now and then, as if to convince her partner that she, too, was pulling.

Emotional Aptitude

Of course, social IQ involves more than knowing how to get another person's attention or teaching the young new information and skills. A deeper social intelligence, one that allows for empathy and grief, is arguably the intangible, almost indescribable fiber that weaves us together. It is tied to our emotional well-being, and as new studies about longevity are showing, these rich social connections are important to our physical health as well. Although we might not be able to ask animals directly how they think and feel about others, we can watch them for clues.

In a 2010 paper psychologist James Anderson of the University of Stirling in Scotland and his colleagues described video footage of an adult female chimp named Rosie and her companions, who lived at the nearby Blair Drummond Safari and Adventure Park, reacting to the death of her elderly mother. Rosie did not eat as much, sleep as soundly or show as much energy as usual for weeks after her mother died, suggesting that the animal may have been in mourning. Other apes, including gorillas in the Democratic Republic of Congo's

Virunga National Park, displayed similar patterns after the death of a parent or a peer.

Grief is not the only complex emotion observed in animals. Psychologist Frans B. M. de Waal of Emory University and his colleagues have shown that chimps prefer actions that will help peers to those that just benefit themselves, indicating that they are inclined toward altruism—a highly social quality once thought to be the exclusive domain of humans.

Social intelligence, as distinctive and intricate as it is, may not be easily separable from other forms of smarts such as problem-solving ability or knowledge of the world. For example, a scrub jay that moves its food stores to keep them secret has performed both social and nonsocial mental gymnastics, including, Lefebvre says, "sophisticated memory mechanisms and time travel [that project] future needs for food." Yet although brilliance very likely spans multiple axes, social cunning is one important—and often underappreciated—element. Indeed, the smarter we humans become about assessing the social intelligence of animals, the more closely related to us they may seem.

Referenced

Pan Thanatology. James R. Anderson, Alasdair Gillies and Louise C. Lock in *Current Biology*, Vol. 20, No. 8, pages R349–R351; April 27, 2010.

Elephants Know When They Need a Helping Trunk in a Cooperative Task. Joshua M. Plotnik et al. in *Proceedings of the National Academy of Sciences USA*, Vol. 108, No. 12, pages 5116–5121; March 22, 2011.

Monkeys Represent Others' Knowledge but Not Their Beliefs. Drew C. W. Marticorena et al. in *Developmental Science*, Vol. 14, No. 6, pages 1406–1416; November 2011.

Theory of Mind in Dogs? Examining Method and Concept. Alexandra Horowitz in *Learning and Behavior*, Vol. 39, No. 4, pages 314–317; December 2011.

About the Author

Katherine Harmon Courage is a contributing editor at Scientific American. *Her first book,* Octopus! The Most Mysterious Creature in the Sea *(Current/Penguin, 2013), examines that animal's famed intelligence, among other astounding traits.*

Insects Recognize Faces Using Processing Mechanism Similar to That of Humans

By Elizabeth A. Tibbetts, Adrian G. Dyer

The wasps and bees buzzing around your garden might seem like simple-minded creatures. They build nests, forage for nectar, raise their young and then die, their lives typically playing out over the course of a single year or less. Some of these species rival humans and other primates in at least one intellectual skill, however: they recognize the individual faces of their peers.

More specifically, members of a species of paper wasp can perceive and memorize one another's unique facial markings and are able to use this information to distinguish individuals during subsequent interactions, much as humans navigate their social environment by learning and remembering the faces of family, friends and colleagues. Further, even certain insects that do not normally memorize faces in the wild can be trained to do so–and can at times even learn to tell human faces apart.

A popular theory of intelligence holds that the exceptionally large human brain evolved to cope with the challenges posed by having to learn and remember many individuals in complex societies. But the discovery that creatures with a brain less than 0.01 percent as large as our own can also identify individuals is forcing scientists to consider how this startling ability evolved and which features of insect brains make facial recognition possible. The answer to the last question, in particular, could help software designers to improve facial-recognition software.

A Fortuitous Discovery

As is often the way with scientific discoveries, the revelation that wasps see one another as individuals resulted from a lucky accident.

As a young graduate student in 2001, one of us (Tibbetts) was working on a project focused on detailing the social lives of *Polistes fuscatus* paper wasps. The project involved painting colored dots on their backs, then videotaping colonies and tracking the interactions among the insects. One day Tibbetts accidentally videotaped a colony with two unmarked wasps. The data would be useless unless she could figure out a way to differentiate between the two insects. As she watched the tape, she suddenly realized that she could tell the unmarked wasps apart by looking at the yellow, brown and black stripes and spots that make up their natural facial markings. Could wasps, she wondered, do the same?

Tibbetts couldn't resist checking. She spent the next few days documenting the fantastic diversity of facial patterns in wasps and then tested whether the creatures could use these patterns as a guide to recognizing individuals. Using an extremely high-tech method—applying modeling paint with toothpicks—she changed a wasp's facial features and then observed the social consequences. Aggression is rare in wasp colonies, so if the wasp was treated more aggressively by nest mates after the makeover, the behavior shift would be proof that wasps pay attention to faces. As a control, she also applied paint to some insects without changing their appearances to rule out the possibility that the wasps were reacting to some aspect of the paint other than its visual effect. She found that the nest mates displayed far more aggression toward the visibly altered wasps than to the control subjects; for the control wasps, interactions with their nest mates proceeded in a business-as-usual fashion. The results showed that wasps do indeed use variation in facial patterns for individual recognition.

Tibbetts was amazed. To appreciate just how astonishing this finding was, consider how humans recognize faces. First, we must perceive a specific arrangement of unique facial features—such as nose, mouth, eyes and ears—and, in our mind, link the arrangement to more abstract information about the person, such as that he or she is our boss or our neighbor. Then we need to quickly recall that pairing each time we see that particular individual.

Interestingly, we learn faces faster and more accurately than we grasp many other types of complex visual information. For instance: if you go to a party, you can quickly memorize the faces of the attendees without any great effort. It would take substantially more time and effort to learn multiple unique, but visually similar complex patterns such as those used in Chinese script. Both faces and Chinese script are composed of multiple elements that come together into a larger whole, but we are better at learning faces than Chinese script because evolution has furnished us with brain adaptations specifically for learning faces. Indeed, in the human brain a region called the fusiform face area is dedicated to face processing. This processing mechanism is so highly specialized that it fails us if an image is simply turned upside down. Likewise, even small modifications to critical regions of a face, such as the eyes, can hamper our ability to recognize a familiar visage.

Although humans as a species excel at face learning, approximately 2 percent of us have some form of deficit. Most face-learning deficits are thought to be hereditary, although they can also emerge in adulthood after injury to the fusiform face area. Given the importance of social recognition in human societies, such disorders can be debilitating. In some extreme cases, people may even have trouble recognizing their spouses and their children, likening the task to trying to learn the identity of different rocks in the garden. In addition, the impaired social development seen in individuals with disorders such as autism may stem at least in part from face-processing deficits.

Given the importance of specialization to human face learning, Tibbetts pondered whether paper wasps could have independently evolved a similar specialization or whether they instead process faces in a different way. To find the answer, she first needed a reliable method for training wasps to attend to images she deemed "correct" while bypassing incorrect ones. Researchers typically train social insects such as honeybees to perform certain tasks by rewarding them with sugar when they make correct choices. Honeybees are eager to work for sugar because collecting food to share with nest mates is one of their primary jobs. Wasps, however, can survive for weeks without eating, so initial efforts to train them using sugar rewards failed. Tibbetts and her then graduate student Michael Sheehan, now at the University of California, Berkeley, eventually found that they could train the creatures to go to the correct image in a choice between two by delivering a small electric shock to the wasps when they selected an incorrect image.

With the new training technique, wasps learned to differentiate between respective pairs of five different types of images: normal wasp face images, simple black-and-white geometric patterns, whole caterpillars (the wasps' natural prey), wasp face images with the antennae digitally removed, or scrambled wasp face images. The wasps rapidly and accurately learned to select specific normal faces within only 20 trials but had greater difficulty learning to discriminate among images in the other four pairings. Most strikingly, simply removing the antennae from a wasp face image or rearranging the face components dramatically reduced their impressive face-learning capacity.

The difference in the wasps' ability to learn normal face images versus antennaeless ones provides very strong evidence that wasps have neural systems specialized for wasp face learning. Antennaeless faces are composed of the same colors and patterns as normal faces, but the wasp visual system does not reliably process and recognize the altered image as a face. The effect of antennae removal on learning indicates that wasps, like people, perceive faces through some kind of holistic processing mechanism. That is, instead of learning each facial feature separately, element by element, the wasp perceives and processes a face as a whole. Thus, for proper learning to occur, the elements must be intact and correctly arranged. The effect of antennae removal on the wasps' ability to learn faces parallels the way human face learning falters when an image of a face is upside down, image brightness is inverted or facial features are scrambled.

The occurrence of face specialization in both humans and wasps suggests that this mechanism could be more widespread in the animal kingdom than initially thought, evolving when social conditions favor it. In the case of *P. fuscatus*, nests are started by groups of queens that work together to survive. But the queens also compete with one another for reproductive dominance. In such situations, it probably pays for cooperating queens to be able to recognize one another and remember how each individual ranks in the dominance hierarchy, hence the evolution of face specialization in this species. By the same token, the face-specialization mechanism should not be present in animals that do not usually need to differentiate individuals.

To test this hypothesis, Tibbetts and Sheehan measured face learning in a close relative of *P. fuscatus*–*Polistes metricus*–which has a different social structure. *P. metricus* nests are typically founded by a solitary queen. With only one queen reproducing, group members have little to gain socially by being able to recognize individual faces. Tibbetts and Sheehan had previously shown that *P. metricus* wasps do not vary in their facial markings and do not naturally recognize individuals. They then posited that *P. metricus* also lacks a special cognitive mechanism for processing faces, unlike the more socially complex *P. fuscatus*. Their findings upheld this supposition.

When given the opportunity, *P. metricus* wasps can learn faces, but it is difficult for them, and they learn faces only about as quickly and accurately as other kinds of images. Furthermore, antennae removal has no effect on the speed or accuracy of their face learning, indicating that this species lacks a specialized holistic-type face-processing mechanism. Instead they process faces as they do any other images: as a collection of independently processed features–perhaps like we humans might learn Chinese script.

How Honeybees See Humans

Given that *P. metricus* can slowly learn faces when trained to do so, despite lacking a specialized mechanism for this activity, one might wonder whether insects with small brains have some capacity to also learn faces of an entirely different type of creature: humans. Inspired by the early paper wasp findings, one of us (Dyer), who studies how bees process visual information, became interested in figuring out whether these insects could learn to tell people apart. In this study, he trained common honeybees to distinguish a target face from a so-called distractor face, presenting them with human visages from a standard neuroscience test. The faces are similar enough that human subjects typically make some errors in judgment. The bees received sweet-tasting sucrose solution for making correct visits to a target face and bitter quinine solution for incorrectly choosing a distractor face. Although it took a while for them to catch on, the bees learned to reliably discriminate the pair of target and distractor faces over the course of 50 trials. The bees also learned to select a target face from a group of novel human faces.

Other experiments using this training procedure revealed some striking similarities between honeybee and human face-processing strategies. First, although the bees learned faces slowly as compared with *P. fuscatus* wasps and humans, they were able to develop some ability to process faces holistically, even though they are not hardwired to do so, as *P. fuscatus* wasps and humans are. Second, honeybees were able to learn multiple viewpoints of the same face and interpolate based on this information to recognize novel presentations. For example, after a bee learns the front and side view of a particular face, it will be able to correctly choose a picture of the same face rotated 30 degrees, even if it has not previously seen this particular image. The honeybees' ability to learn faces is particularly unexpected because honeybee society is far simpler than that of the wasps, consisting of a single queen and a mass of essentially identical workers that do the same tasks. Honeybees do not have distinctive facial markings, and their interactions in the hive often rely on complex pheromonal signals rather than visual cues.

The results suggest that this line of research could be a boon to efforts to develop automatic face-recognition systems. Identifying the same face from different angles is often thought to be a difficult task and is one of the major challenges of machine face recognition. But the tiny brains of these bees are much simpler than primate brains, so once we figure out what processing tricks the bees use for solving such complex problems, we should be able to apply them relatively simply to machine vision to enhance face-recognition software.

Taken together, these studies in insects tell us something fundamental about how face recognition may have evolved. The simple system that *P. metricus* wasps and honeybees have that allows them to learn to recognize faces through experience even though they do not normally distinguish among individuals in their everyday lives may be based on the general pattern-recognition abilities these species use during foraging. It may also function as an intermediate step in the evolution of face specialization. When the forerunners of today's *P. fuscatus* wasps found themselves in a novel social environment in which telling individuals apart aided their survival and reproduction, the wasps could learn to identify individuals. Over time natural selection most likely built on that foundation, changing their brain to produce the face specialization that enhanced the wasps' ability to sort friend from foe quickly and reliably. The intermediate system allowed this adaptation to evolve rapidly: *P. fuscatus* and *P. metricus* are closely related, and their last common ancestor would have had the comparatively primitive face-learning system seen in *P. metricus*. Thus, *P. fuscatus*'s biological adaptation to processing faces efficiently must have evolved recently, after its lineage diverged from that of *P. metricus*.

So the next time you are out in your garden, take a minute to appreciate the resident wasps and bees. There is ever so much more going on in their teensy brains than we could have imagined possible.

Referenced

Visual Signals of Individual Identity in the Wasp *Polistes fuscatus*. Elizabeth A. Tibbetts in *Proceedings of the Royal Society of London B*, Vol. 269, pages 1423–1428; July 22, 2002.

Individual Recognition: It Is Good to Be Different. Elizabeth A. Tibbetts and James Dale in Trends in *Ecology and Evolution*, Vol. 22, pages 529–537; October 1, 2007.

Insect Brains Use Image Interpolation Mechanisms to Recognise Rotated Objects. Adrian G. Dyer and Quoc C. Vuong in *PLOS ONE*, Vol. 3, No. 12, Article No. e4086; December 31, 2008.

Configural Processing Enables Discrimination and Categorization of Face-like Stimuli in Honeybees. Aurore Avargues-Weber, Geoffrey Portelli, Julie Benard, Adrian G. Dyer and Martin Giurfa in *Journal of Experimental Biology*, Vol. 213, pages 593–601; February 15, 2010.

Specialized Face Learning Is Associated with Individual Recognition in Paper Wasps. Michael J. Sheehan and Elizabeth A. Tibbetts in *Science*, Vol. 334, pages 1272–1275; December 2, 2011.

About the Authors

Elizabeth A. Tibbetts is an associate professor at the University of Michigan. She studies how evolution shapes animal behavior and cognition.

Adrian G. Dyer is an associate professor at RMIT University in Melbourne. He conducts behavioral studies to investigate how the visual systems in different animals process complex information.

A Famed Dolphin-Human Fishing Partnership Is in Danger of Disappearing

By Rachel Nuwer

People in Laguna proudly refer to their southern Brazilian city as the "national capital of fish-herding dolphins." For at least 140 years artisanal fishers and bottlenose dolphins have worked together in careful synchrony to catch mullet in a local lagoon. The spectacle of nets flying through the air while dolphins dive into the murky water has become a popular attraction for tourists, and it is recognized by local authorities as an intangible cultural heritage.

In 1998 scientists confirmed that this renowned example of human-wildlife cooperation aids at least one of the parties: the fishers, who enjoy larger catches when they join forces with dolphins. Most believed that the dolphins were reaping rewards, too, but this hypothesis was more difficult to test.

Now researchers have finally documented that the benefits are indeed mutual. An exhaustive new study published in the *Proceedings of the National Academy of Sciences USA* shows that dolphins that team up with fishers gain more food and have an edge in survival compared with those hunting without human partners.

Laguna's resident population of up to 60 dolphins uses echolocation to find schools of mullet that humans are unable to detect in the opaque water. Then the cetaceans herd a school toward the fishers, who are typically standing in the shallows just offshore. The dolphins give a cue, such as a sudden deep dive, to let the fishers know when to cast their nets; the net casting in turn breaks up the school of speedy fish, making it easier for dolphins to capture individual mullet.

"Each [species] brings a new skill to the table that increases their mutual success," says Mauricio Cantor, the study's lead author and a behavioral ecologist at Oregon State University. The research indicates, however, that these celebrated interspecies hunts are in danger of vanishing. "Our data suggest this interaction is becoming rarer over time," Cantor says. "If things continue the way they are, these interactions might disappear in the next 50 to 60 years."

Laguna's cooperative fishers and dolphins are one of the few remaining examples of a millennia-old tradition across the world. Fossil evidence from Europe indicates that humans and wolves might have collaborated to hunt prey as early as 32,000 years ago. In the first century C.E. Pliny the Elder mentioned fishers working with dolphins in what is now southern France. Since then, seemingly cooperative interactions with people have been recorded in at least 16 species–mostly cetaceans but also some birds–in countries such as Australia, Myanmar, Mauritania and Japan.

Many such collaborations have ended, however. And those left are almost all in decline because of pollution, overfishing and habitat loss, combined with the general disconnection of humans from the natural environment. "Today most interactions we have with wildlife tend to be antagonistic and not mutually beneficial," Cantor says. "So it's really important to understand how these things happen and what we can do to safeguard this unique biological phenomenon."

To analyze Laguna's human-dolphin cooperation, Cantor and his colleagues collected data using drones, underwater microphones, sonar cameras, GPS and interviews with fishers. The study's senior author, Fábio Daura-Jorge of the Federal University of Santa Catarina in Brazil, also provided 15 years' worth of survey responses from fishers, as well as photographs and dolphin behavior records for both individual animals and groups.

After pulling it all together, the researchers found that fishers were 17 times more likely to catch mullet when they worked with dolphins than they were on their own, and dolphins had more success in these partnerships as well (the researchers are still working to quantify the dolphins' gains). Dolphins that partnered with fishers also experienced a 13 percent boost in survival, most likely because they spent most of their time in the lagoon. In areas nearby, some fishers use illegal drift nets that drown dolphins as bycatch. "By avoiding these encounters, cooperative dolphins have a higher probability of surviving," Cantor says.

These findings support "the long-standing hypothesis that these dolphins experience a higher prey capture rate when interacting with fishers," says University of Cape Town behaviorial ecologist Jessica van der Wal, who was not involved in the research. "I applaud the authors for their comprehensive and creative angles to better our understanding of the natural history of [this] complex cooperation."

The team also built a model to predict how fisher-dolphin interactions might change over time. Fishers have anecdotally reported a decline in joint hunting practices over the past 15 years, and the model suggests that this trend will continue, putting the practice in danger of extinction. The main drivers for that decline, the authors found, are decreasing mullet populations caused by overfishing and climate change, along with livelihood changes in the artisanal fishing community.

The scientists propose some solutions to help prevent this loss: Fishers, for example, could be incentivized to keep working with dolphins if buyers of their catch were willing to pay a premium for their fish. This way, Cantor says, "even if catch declines over time, there will be a higher reward for fishers in monetary terms."

Protecting dolphins from becoming bycatch would also help. The government could invest in more policing to remove illegal nets and apprehend violators. And raising awareness is key. "Getting people excited about these rare phenomena is a way to add more value to these kinds of cultural practices that seem to be fading away," Cantor says. "Preserving cultural diversity has been shown to promote the preservation of biological diversity as well."

The "exemplary" new study is "of huge importance" for showing both how this cooperation is achieved and how it can be conserved, says University of Bristol behavioral biologist Stephanie King, who was not involved in the research. "Humans are renowned for the ways they cooperate," she says, "but what is even more remarkable is two distantly related species with very different evolutionary histories acting cooperatively to achieve a common goal."

About the Author

Rachel Nuwer is a freelance science journalist and author who regularly contributes to Scientific American, *the* New York Times *and* National Geographic, *among other publications. Follow Nuwer on Twitter @RachelNuwer.*

Section 5: Extreme Animal Evolution

Two-Headed Worms Tell Us Something Fascinating about Evolution

By Ashleigh Papp

Researchers looked back at more than 100 years of research and found that a fascination with annelids with mixed up appendages was strong—and that research still has relevance today.

Ashleigh Papp: This is *Scientific American's* 60-Second Science, I'm Ashleigh Papp.

You might say that Guillermo Ponz is a scientific monster hunter–even though he thinks that term, "monster," never really captured his subjects right.

Guillermo Ponz: So they're regular animals that have gone through different developmental processes that would end up building a body, that is not what you expect.

Papp: What this researcher based in Madrid, Spain, actually loves is the oddly amazing animals. After all, he studies *two-headed worms*.

Ponz: We have these worms that are usually regular worms like with one head and one tail, that's normal, but sometimes they may have two heads or two tails. And on the other side, there are worms, which have one head and many tails always.

Papp: Officially, he looks at bifurcated annelids, meaning things like earthworms that have come out of their larval stage with two heads, or spontaneously sprouted two tails, or ... some other combination of mixed up appendages.

We know that certain species, like some salamanders and insects, have the ability to regrow appendages in a time of need. But there's this one phylum of worms, the annelids, that can re-grow unlike anything else that we've ever seen in the kingdom.

Their segmented bodies, like an earthworm with rows of ringed compartments, help them easily regrow a new head or tail at the first sign of trouble.

Or even crazier, they can regrow an entirely new right side of their body if sliced in half.

Ponz: ... worms that do these crazy things that are very weird, very, you know, very, very strange things that these worms should not, quote-unquote, should not do.

Papp: Once Ponz started studying the anatomically death-defying lengths to which these worms would go to grow and survive, he was totally pulled in.

And he realized that he and his team weren't the first to be fascinated. Ponz found that there was a golden age of research on "monster creatures" during the 18th and 19th centuries.

Ponz: ... 100 plus year old literature would refer to the sentiments of monsters, creatures, or monsters, or oddities or, you know, they're, they're all these these variations that describe them. And in the end, these, these animals are not, not monsters.

Papp: A fixation with the "reanimated Monster" makes sense, especially back then. Author Mary Shelley's novel, *Frankenstein*, was published in 1818, and it only further intensified interest–and some of that interest translated into actual research.

Ponz and a team of international researchers conducted a wide-sweeping review of the existing knowledge about monster worms. They dove into 275 years' worth of research–combing through scientists' observational journals, reading historical texts, and even reaching out to the broader scientific community to see if anyone knew anything about records of abnormal worms.

They wanted to understand all of the different types and patterns of bifurcation and see if there were any clues about how the oddities developed.

Their search landed them in a jackpot of both history and science.

They came across documents and drawings of bifurcated worms from around the world–in Latin, French and German, all the way to Russian, Japanese, and even Indonesian. All in all, they spent over a year working through the archives, translating old texts, and following the trail of monster worm clues.

What they learned is that bifurcation in worms has been observed in over 60 species of worms across the annelid family tree, and in some species, up to 20% of the juveniles ended up with some form of bifurcation. This work was recently published in the journal *Biological Reviews*.

Ponz: And that means, for example, in the in the case of bifurcation, that when an animal is cut, and is regenerating, for example, the tail, there needs to be some mechanism that specifies where this tail is going to be, how it's going to be oriented, what is posterior, what is anterior, what is left, what is right, what is dorsal, what is ventral. And these mechanisms can be disturbed. And these might lead to different anatomies.

And that gives us clues about what is important during this process. Of course, these are the baby steps. So we are just pointing towards this process, this phenomenon, we are saying, Okay, hello, this happens, there are these animals that are doing these weird things. We should not forget about them, let's look into them.

Papp: They also realized that there's a strong correlation between the type of bifurcation and the internal organ development. Meaning, the way that the worms were split reliably indicated if extra sets of organs were present.

With this type of intel, Ponz and his team were able to essentially draw up a blueprint, or how-to guide, for reliably and repeatedly creating bifurcated worms ... which is potentially a very useful resource for scientists interested in studying the mechanisms of development.

This long-forgotten study of worm developmental anomalies seems poised for a comeback. According to Ponz, this information could extend far beyond the annelid and even insect worlds to help us better understand how things like growth and development actually happen ... in both the normal and the monster ways.

Ponz: In a sense, we are now following this trend that they started then, studying these animals to try to understand bigger pictures in nature. Usually development leads to a certain way to assert them to a certain point. So you have a development that ends up in anatomy that's more or less conserved. But sometimes it doesn't. And that can teach us something about development processes. And that's interesting.

Papp: For *Scientific American*'s 60-Second Science, I'm Ashleigh Papp.

[The above text is a transcript of this podcast.]

Killer Whales Are Speciating Right in Front of Us

By Rüdiger Riesch

Just offshore from the pebble beaches of Bere Point on Malcolm Island, British Columbia, the *Naiad Explorer* rocks gently in the waters of the Queen Charlotte Strait. The sun has burned off most of the morning mist, save for a thin layer that still shrouds the tips of the island's cedars, firs and spruces. I watch from the boat as three killer whale brothers named Cracroft, Plumper and Kaikash gently scrape their bodies against the small, smooth stones in the shallows off the bow. The brothers have already spent the better part of an hour here absorbed in this activity. Soon they will leave to hunt for salmon or look for mates.

Exactly why the creatures engage in this scraping behavior, known as beach rubbing, is uncertain. Most experts assume that it aids in sloughing off dead skin and dislodging external parasites, but it might also be for pleasure. Whatever the motivation behind it, beach rubbing, though rarely observed in other cetaceans–the group that includes whales, dolphins and porpoises–is commonplace here. It is part of the distinctive cultural fabric of the northern resident killer whales, a community that claims the waters around northern Vancouver Island as home during the summer months. (Despite their name, killer whales are actually large dolphins.)

The northern resident killer whales are not the only ones with unique behaviors. Observations made since the 1970s have shown that killer whale populations around the globe each have unique ways of doing everything from hunting to communicating. Physical traits, including coloration, body size and dorsal fin shape, vary among groups as well, albeit somewhat more modestly. These cultural and physical differences, along with the astonishing degree of genetic diversity documented in these creatures over the past 15 or so years, suggest to me and many other researchers that today's killer whale populations, rather than simply representing the single species scientists have long envisioned, are actually in the midst of going their separate evolutionary ways. That is, they appear to be splitting into new species that, if this process continues, ultimately will be unable to produce viable or fertile offspring with one another.

Intriguingly, their cultural differences may be driving this diversification: the whales seem to mate with individuals that largely share their customs, to the exclusion of those that do not, a preference that creates the conditions needed for speciation to occur. If so, then killer whales could provide a striking example of a speciation mechanism not considered in the classical theory of how new species arise. They might also offer insights into how another group of creatures–*Homo sapiens* and our extinct predecessors–diverged into an array of species that once shared the planet.

A Diversity of Killers

For more than a century biologists have looked to geography to explain how speciation occurs. In the favored scenario, called allopatric speciation, two populations of an ancestral species become physically separated from each other, often by a geographical barrier of some kind–perhaps a mountain range, desert or river–that prevents the populations from interbreeding. If this separation persists long enough, over time each population will follow its own independent evolutionary trajectory, acquiring different genes that may help them survive in different environmental conditions, for example, or that may accumulate randomly through a process known as genetic drift. Eventually, so the theory goes, the two populations can become so genetically different from each other that if they come into contact again, they cannot interbreed successfully.

Overwhelming evidence from numerous organisms ranging from species of *Alpheus* snapping shrimp that live on either side of the Isthmus of Panama to *Cyprinodon* pupfish species found only in isolated springs in California and Nevada shows that geographical isolation does indeed facilitate the speciation process. Yet sometimes two or more subpopulations of dissimilar appearance will emerge within the same geographical area and ultimately differentiate into separate species. Scientists, including famed German evolutionary biologist Ernst Mayr, traditionally held that at least some period of isolation was essential to the process of speciation and that speciation entirely within the same area was therefore either impossible or at least very rare. More recent work has demonstrated that geographical isolation in the traditional sense is not always necessary for speciation to occur, however.

Indeed, biologists now largely accept that certain species, among them the astonishingly diverse cichlid fishes found in the crater lakes of East Africa and Nicaragua, as well as the *Howea* palm trees found on Lord Howe Island in the Pacific, have evolved in the absence of such isolation. In the parlance of biologists, they have undergone sympatric speciation, and in the case of cichlids, this was apparently mostly driven by different cichlids adapting to exploit different food sources (but without the influence of culture), whereas different Lord Howe palms have evolved different flowering times. Documented examples of sympatric speciation among mammals are rare, though, which makes the case of the killer whales especially interesting.

Killer whales (also called orcas and, in some regions, blackfish) are the most widely distributed mammal on earth after humans. They inhabit all the world's oceans and can travel more than 100 kilometers a day or upward of several thousand kilometers within just a few weeks. There are no known geographical barriers that would prevent individuals from one population from mingling with members of neighboring populations. Yet scientists have now shown that in various marine areas, several ecologically distinct forms, or ecotypes, of killer whales live side by side without fraternizing. One type, for instance, might live mainly on a particular kind of fish, whereas another type might prefer seals.

The best-studied assemblage of killer whale ecotypes lives in the Northeast Pacific. There research begun in the early 1970s by Canadian scientist Michael Bigg led to many remarkable discoveries. First, he noticed that individual killer whales differed in the shape and size of their dorsal fins and in the shape, size and coloration of the so-called saddle patch–the grayish white area behind the dorsal fin. Biologists can use those traits to identify individual killer whales, just as forensic scientists use facial features and fingerprints to identify individual humans. Second, Bigg and his colleagues, including John K. B. Ford and Graeme M. Ellis, both at Fisheries and Oceans Canada, and Kenneth C. Balcomb III of the Center for Whale Research in Washington State, found that three different killer whale ecotypes coexist in the waters off the West Coast of the U.S. and Canada: the so-called resident, transient and offshore forms. Resident killer whales specialize in foraging for fish, especially salmon; transient killer whales target marine mammals and the occasional seabird; and offshore killer whales seem to specialize in a second kind of fish diet that includes Pacific halibut and Pacific sleeper sharks, although their habits remain largely mysterious because encounters with them are rare.

In recent years research led by Olga A. Filatova of Moscow State University, Alexander M. Burdin of the Russian Academy of Sciences, and Erich Hoyt of Whale and Dolphin Conservation in England has revealed that residentlike and transientlike ecotypes also share the Russian waters of the Northwest Pacific around the Kamchatka Peninsula. Thus, we now know that there is a more or less continuous band of resident populations coexisting with transient populations that connect the Northeast Pacific with the Northwest Pacific via the Aleutian Islands.

Halfway across the world, killer whale populations from around Iceland, Shetland and Norway in the Northeast Atlantic have their own food preferences. Scientists, including Volker B. Deecke of the University of Cumbria in England, Andrew D. Foote of the University of Bern in Switzerland and their colleagues, have reported on two groups: type 1 killer whale populations forage for fish, in particular herring and mackerel, and type 2 killer whales pursue seals. More research is needed to fully understand the dietary differences between the groups.

The Southern Hemisphere hosts geographically overlapping ecotypes, too. John W. Durban and Robert L. Pitman, both at the National Oceanic and Atmospheric Administration, and their colleagues have discovered at least four different ones in Antarctic and sub-Antarctic waters. "Type A" killer whales appear to specialize in hunting Antarctic minke whales, "type B" killer whales, on the other hand, come in a large form (the so-called pack ice killer whale) that focuses on seals, and a small form (the Gerlache killer whale) that favors penguins. "Type C"–the smallest known killer whale–hunts Antarctic toothfish. For its part "type D," like the offshore killer whales of the North Pacific, seems to be an open ocean ecotype that remains largely elusive. This ecotype is known to eat Patagonian toothfish from longline fisheries, but it probably dines on other prey as well.

Once scientists realized that all these factions existed within the killer whales, they began to wonder about the origins of these groups. Were the populations already living in the same geographical region when they started to differentiate, or did they start diverging at a time when they lived apart and only later colonized the same region after they had already started down separate evolutionary paths? Current evidence is mostly inconclusive for many Northern Hemisphere killer whales. Whereas several studies by Foote and his colleagues suggest that killer whale divergence in the North Pacific happened while populations were geographically isolated (allopatric speciation), other analyses by Alan Rus Hoelzel of Durham University in England and his collaborators hint that these ecotypes might have always co-existed (sympatric speciation). For the killer whales of the Antarctic, though, the evidence is clearer: most, if not all, of these killer whale ecotypes probably diverged sympatrically while living in the same geographical region.

However the now sympatric ecotypes initially began to diverge, they did so rapidly. In May of this year, Foote and his colleagues reported that their genomic analysis of five killer whale ecotypes from the North Pacific and the Antarctic showed that these forms evolved from a common ancestor within the past 250,000 years.

What is keeping them separate now? Breeding records maintained and occasionally published by SeaWorld, which houses killer whales originally captured in both the North Atlantic and the North Pacific, demonstrate that matings between different killer whale ecotypes produce viable and fertile offspring, in contrast to the usually infertile mules and hinnies that result from hybridization between horses and donkeys. It is therefore extremely unlikely that genetic incompatibilities between killer whale ecotypes are preventing interbreeding in natural populations. Rather mounting evidence suggests that cultural differences are to blame.

Culture Clash

Like many organisms that have undergone speciation, killer whales are diversifying by exploiting different food sources and evolving various traits that presumably help them get those foods. Some of these distinguishing features are physical, such as the generally larger, stronger build of the mammal-hunting killer whales. But the most dramatic specializations have occurred in cultural behaviors related to food acquisition. Because these and other behaviors are found only in certain populations where they appear to be passed on between members of the same generation and from generation to generation through social interactions (so-called social learning), rather than being innate, biologists consider them to be cultural.

For example, populations of killer whales that hunt marine mammals have learned to intentionally strand themselves to capture inexperienced sea lion and elephant seal pups right off the beach. Scientists have observed this behavior in two groups of killer whales that are distinct from the previously mentioned ecotypes. One inhabits the waters around the Crozet Archipelago in the Indian Ocean between Africa and Antarctica; the other dwells near the Peninsula Valdés on Argentina's Atlantic Coast. Apparently both populations invented this hunting strategy independently in response to their prey choices and the physical characteristics of their hunting grounds, where deep water channels and river outlets allow the killer whales to stay largely submerged until they are just meters away from their quarry.

In Antarctica, the large, pack ice form of the type B killer whales have invented another ingenious hunting strategy to gain access to seals: wave washing. The seals often haul out on small ice floats, where they feel safe from predators. But the ice-pack killer whales have learned to create waves that wash the seals over the ice float and into the water, where they are easier to nab.

Type 1, fish-eating killer whales around Iceland and Norway, for their part, have developed an entirely different strategy–dubbed carousel feeding–to hunt the herring that form the mainstay of their diet. A pod of the killer whales will herd a school of herring into a tight ball close to the water surface, where the fish cannot escape into the depths. Then individual pod members will swim right into the ball and slam their tail fins into the fish to debilitate and kill them.

Killer whales even communicate differently depending on what they eat. Indeed, it is in their acoustic communication signals that the most astonishing cultural diversity is found. Like other dolphins, they use three different acoustic signals: echolocation clicks, which are used to navigate and to locate prey, and pulsed calls and whistles, both of which are used for communication with their compatriots. Not only do the pulsed calls and whistles differ among killer whales from different geographical regions, but they also differ among populations that inhabit the same region.

The reason for this intraregional variation in signal production and use becomes clear when one considers the different challenges the ecotypes face. Killer whales that hunt marine mammals, for example, must contend with the excellent underwater hearing ability of their quarry: eavesdropping prey can use any sound the killer whales produce to detect them and subsequently evade capture. Transient killer whales in the Northeast Pacific and mammal-hunting killer whales in the North Atlantic thus use acoustic signals only very sparingly; most of the time they swim and hunt in stealth mode. Fish-eating killer whales do not have the same problem, so they are typically very chatty with one another, and they do not skimp on echolocation when navigating and tracking prey.

Furthermore, many pulsed calls and, as some of my own research has shown, some whistles are highly stereotyped. That is, the signals can be further differentiated into discrete sounds, like letters in an alphabet. (There is no evidence to suggest that killer whales use these signals in any way that really resembles our human use of words and sentences, however. Rather the context within which a signal is used seems to provide the meaning.) These discrete sounds exhibit geographical variation and ecotype variation. But they also often vary among social groups within an ecotype. For instance, among northern resident killer whales–a population of fish-eating killer whales that inhabits the waters from approximately around the middle of Vancouver Island up into southeastern Alaska–each family unit has its own repertoire of seven to 17 discrete calls. Killer whale families that share a portion of their dialects are grouped together into acoustical clans: A-clan, G-clan and R-clan for the northern resident killer whales.

The different discrete call types and family dialects are so distinctive that those of us who work on these killer whale populations can assign individuals into the correct ecotype, clan (for northern resident killer whales) or even family unit, based on recordings of their discrete call repertoire alone. These differences figure importantly in mate choice. Genetic analyses of the northern residents by Lance Barrett-Lennard of the Vancouver Aquarium Marine Science Center have shown that call similarity largely mirrors genetic similarity. Most matings take place between members of different clans, which have correspondingly different calls. The finding implies that northern residents find other northern residents that sound different from themselves more attractive than those that sound similar. Thus, the dialects offer a nifty way to prevent inbreeding.

That killer whales have all these ecotype-specific customs and appear to dislike socializing and mating with foreigners from other ecotypes despite being biologically capable of doing so suggests that culture is keeping these ecotypes apart. Eventually, if this separation persists for enough generations, then these different ecotypes might evolve additional differences in their DNA that could render them genetically incompatible. Culture in killer whales thus has the potential to take the place of geographical isolation in facilitating speciation by preventing mixing between populations.

The killer whale findings raise interesting questions about diversification within the human family. Traditionally anthropologists thought that most selective pressures that shaped our evolution were the result of changes happening purely in our external environment. But recent genetic analyses indicate that a large part of our evolution might have resulted from certain sometimes very locally restricted, cultural innovations. The practice of cattle farming has driven the evolution of lactose tolerance in certain European and African populations; the high-fat diet of Inuit people in Greenland has driven the evolution of a more efficient fat metabolism in that population. Although all modern human populations clearly belong to the same species and mix routinely with one another, for most of human prehistory, multiple human species shared the planet. Might culture have also played a role in driving speciation among those early members of the human family?

Separate Ways

Despite the amazing advances in decoding how killer whales have diversified, scientists still have much to learn. Do other areas in less researched regions of the world also host sympatric killer whale ecotypes? Some preliminary studies hint that the oceans around Africa might; those around South America and southern Asia come to mind, too. Also, what are the communications systems of Antarctic and sub-Antarctic killer whales like, and what are their social structures? With the advent of modern tools for studying genomes, the future of speciation research in general–and killer whale research in particular– looks very bright. Perhaps in the not so distant future these and other new technologies will enable us to unequivocally determine what the geographical arrangements of killer whale populations were during all phases of their diversification.

Already we know that culture can divide killer whale populations that live side by side. Maybe a few years from now biologists will recognize these ecotypes as different species, each restricted to a certain geographical area of our oceans, each with its own very specific diet and customs, each with the potential to diverge and form yet more new twigs on the tree of life.

Referenced

Cultural Traditions and the Evolution of Reproductive Isolation: Ecological Speciation in Killer Whales? R. Riesch et al. in *Biological Journal of the Linnean Society*, Vol. 106, No. 1, pages 1–17; May 2012.

Phylogenomics of the Killer Whale Indicates Ecotype Divergence in Sympatry. A. E. Moura et al. in *Heredity*, Vol. 114, No. 1, pages 48–55; January 2015.

Geographic and Temporal Dynamics of a Global Radiation and Diversification in the Killer Whale. P. A. Morin et al. in *Molecular Ecology*, Vol. 24, No. 15, pages 3964–3979; August 2015.

About the Author

Rüdiger Riesch is a lecturer in evolutionary biology at Royal Holloway, University of London. His research focuses on the mechanisms that create, maintain and constrain biodiversity, with a special emphasis on speciation that occurs as a result of a population exploiting a new ecological niche.

'Chatty Turtles' Flip the Script on the Evolutionary Origins of Vocalization in Animals

By Pakinam Amer

Recordings of more than 50 species of turtles and other animals help scientists reassess the origins of acoustic communication in vertebrates.

Pakinam Amer: This is *Scientific American*'s 60-Second Science. I'm Pakinam Amer.

Clicks, clucks, grunts and snorts–these are not sounds that we typically associate with turtles.

[CLIP: Audio of South American juvenile turtles]

Amer: They're actually thought to be very quiet or even silent. But it looks like we may have grossly underestimated how much sound they can make. Now a new study in *Nature Communications* has collected vocal recordings from 53 species of turtles and other animals that were otherwise considered to be mute.

[CLIP: Audio of South American juvenile turtles]

Amer: Those clicks you've just heard were calls made by baby giant Amazon River turtles swimming together. A group of evolutionary biologists and other scientists in five different countries pored over these recordings and combined them with vocal repertoires of about 1,800 animal species from other studies.

Amer: They were able to piece together evidence that the last common ancestor of all lungfish and tetrapods started vocalizing more than 400 million years ago. (And just in case you aren't familiar, tetrapods are four-limbed vertebrates that include amphibians, mammals, birds and reptiles.) That's at least 100 million years earlier than previous studies had suggested.

Amer: The new revelations amount to a rewriting of the acoustic history of animals with backbones.

Gabriel Jorgewich-Cohen: I did fieldwork in the Brazilian Amazon with a researcher that published one of these first papers showing that turtles can communicate acoustically, and that inspired me. So I went back home, and I got a piece of equipment, and I started recording my own pets. And I discovered that they were producing sounds as well, and the species I had were not known to produce sounds. So I started thinking maybe they all do, and I went out there, and I recorded as many as I could [*laughs*].

Amer: That was Gabriel Jorgewich-Cohen, a researcher at the University of Zurich and study co-author. By the way, the pets he's talking about are giant Amazon River turtles, more commonly known as red-eared slider turtles in the U.S.

Jorgewich-Cohen: This is the only species known to have post-hatch parental care among all turtles, which is pretty amazing. And they discovered this by recording the sounds of the animal–not only this species but also sea turtles, for example. When they are in the nest, the hatchlings start vocalizing from within the egg to synchronize hatch. And also when they come out altogether, they individually have less chance of being eaten by another animal. And in the case of the Amazon River turtle, when they go to the water, the females are there, waiting for them, and they are also vocalizing. And they find each other, and then they migrate together up the river to the forest.

Amer: A previous study published in 2020 by researchers at the University of Arizona concluded that only two of 14 families of turtles vocalized. It also stated that acoustic communication evolved independently in most major tetrapod groups, with origins in the range of 100 million to 200 million years ago. But now we know that's not the case.

Jorgewich-Cohen: I was very surprised–happily surprised–when I found so many different types of sounds. And I kept recording more and more animals. And every animal I recorded made sounds; I had no negative results whatsoever. And that was surprising by itself.

Amer: Jorgewich-Cohen recorded hundreds of hours' worth of footage over two years–not just of turtles but also of lungfish, tuatara and other creatures. Animals typically produce sounds for many reasons: to define territory, to attract a mate or to communicate with their young ones. It's a useful skill.

Jorgewich-Cohen: I found that for many turtle species, there are sounds that are only made by males, there are some that are only made by females, and some only by juveniles, and some that males will only make when they are in front of the female.

Amer: If there's one animal from this study that I would've sworn is 100 percent mute, it's the caecilian. For those who're not familiar, let me paint a little picture: Caecilians are slippery, slimy and slithery little things. They burrow, and they look like earthworms or even snakes. But they're neither. They're in fact amphibians. They have a backbone and a skull, jaws and all, but no limbs. And like many tetrapods, they emit sounds through their respiratory tract, just like their common ancestor. It's actually not very easy to come across one.

Jorgewich-Cohen: The caecilian was a special one because I definitely expected it not to make any sounds. And it's not only that it does, but it makes very strange and very loud sounds.

[CLIP: Audio of caecilain]

Amer: Not to be crass, but that sounds a bit like a fart.

Jorgewich-Cohen: When I heard it for the first time, I started laughing, and I sent it to my friends who did fieldwork with me. They also started laughing, and they said, "I cannot believe you. You made the sound with your mouth, and you're sending me the file." I was like, "No, I swear."

Amer: The study, "Common Evolutionary Origin of Acoustic Communication in Choanate Vertebrates," is less focused on the function of these sounds and more on the evolution of acoustic signals. But in future studies, the researchers plan to dig deeper by analyzing the sounds further in an attempt to understand what they mean.

Jorgewich-Cohen: We try to also make footage of the animals while we're recording the sounds so we could try to correlate any type of behavior to the sound that they were making and try to understand how they use the sounds or what ideas they convey.

Amer: Sometimes Jorgewich-Cohen and his colleagues would find more than 30 different sounds in a single species' repertoire. It seems that the more socialized the animal is, the more vocally diverse it is, he says. But further studies are needed to confirm this.

Jorgewich-Cohen: Hopefully this is the beginning of a new field of study. So people are going to go out there and try to record more of these animals and get to new conclusions and new discoveries. But it will be really cool if we could, for example, do playback experiments and try to understand if they reply to the sounds we make. And then we can start understanding what these sounds mean and how they are used.

Amer: Thank you for listening! For *Scientific American*'s 60-Second Science, I'm Pakinam Amer.

[The above text is a transcript of this podcast.]

GLOSSARY

acclimate To adjust physiologically or behaviorally to a new climate or circumstances.

austral Related to the Southern Hemisphere.

axon A fiber that extends from a nerve cell such as a neuron to carry electrical impulses away from the cell.

bifurcation Dividing into two branches.

detritus Debris or waste.

ecotype A specific race or type of a species that occupies a particular habitat or environment.

foraging Searching for food or other supplies.

fraternizing Associating with or developing a friendship.

inertia The property of remaining unchanged unless acted on by an external force.

locomotive Concerning the ability to move from one place to another.

longevity Long life.

resilience The ability to survive or withstand difficult circumstances or events.

speciation The process of evolutionarily developing into a new and distinct species.

symbiotic A mutually beneficial relationship or interaction between organisms.

torpor A state of lower activity that some animals experience in the winter.

FURTHER INFORMATION

Arnold, Carrie. "Dolphin Moms Use 'Baby Talk' with Their Calves, Rare among Non-Human Species," *National Geographic*, June 26, 2023, https://www.nationalgeographic.com/animals/article/bottlenose-dolphins-mothers-baby-talk.

Bittel, Jason. "'The Perfect Revenge?' Birds Are Building Fortresses from Anti-Bird Spikes," *National Geographic Magazine*, July 17, 2023, https://www.nationalgeographic.com/animals/article/bird-spikes-europe-magpies-nests.

Bryce, Emma. "What Is the Biggest Spider in the World?," *LiveScience*, August 4, 2023, https://www.livescience.com/animals/spiders/what-is-the-biggest-spider-in-the-world.

Freedman, Ethan. "What Is the Deepest Diving Mammal?," *LiveScience*, August 6, 2023, https://www.livescience.com/animals/whales/what-is-the-deepest-diving-mammal.

Incorvaia, Darren. "New Antarctic Starfish Are Doting Parents and Vicious Predators," *New York Times*, August 4, 2023, https://www.nytimes.com/2023/08/04/science/antarctica-starfish-brooding.html.

Kwan, Jacklin, and *LiveScience*. "Dolphins and Whales Will Never Evolve Back into Land Animals," *Scientific American*, July 17, 2023, https://www.scientificamerican.com/article/dolphins-and-whales-will-never-evolve-back-into-land-animals/.

Oza, Anil, and *Nature* magazine. "'Virgin Birth' Engineered into Female Animals for First Time," *Scientific American*, August 1, 2023, https://www.scientificamerican.com/article/virgin-birth-engineered-into-female-animals-for-first-time/.

Roth, Annie. "These Clever Fish Have Mastered the Fakeout," *New York Times*, August 7, 2023, https://www.nytimes.com/2023/08/07/science/trumpetfish-hunting-stalking-horse.html.

CITATIONS

1.1 50,000 Worms Tangled Up in a Ball Unravel in an Explosive Burst when a Predator Appears by Jack Tamisiea (May 10, 2023); 1.2 Tiny Spider Fells Prey Many Times Its Size by Jack Tamisiea (March 16, 2023); 1.3 Social Animals Seek Power in Surprisingly Complex Ways by Lee Alan Dugatkin (April 15, 2022); 1.4 The Surprising Architecture in Bees' Honeycombs by Saugat Bolakhe (November 1, 2021); 1.5 Play Is Serious Business for Elephants by Caitlin O'Connell (August 1, 2021); 1.6 How This Zombie Fungus Turns Cicadas into Horror-Movie Sex Bots by Rachel Nuwer (May 27, 2021); 2.1 Octopuses Redesign Their Own Brain When They Get Chilly by Rachel Nuwer (June 8, 2023); 2.2 How Do Birds Know When to Migrate? by Meghan Bartels (April 20, 2023); 2.3 How Bears Hibernate without Getting Blood Clots by Jack Tamisiea (April 13, 2023); 2.4 Giraffes vs. Blue Whales vs. Dinosaurs: Which One Builds Its Nervous System Fastest by Douglas H. Smith, Jeffrey M. Rodgers, Jean-Pierre Dollé, Mathew J. Wedel (September 29, 2022); 2.5 How Snakes Breathe while Crushing Prey by Lars Fischer, Joanna Thompson (July 1, 2022); 2.6 The Surprising Secret of Snakes' Venomous Bites by Riley Black (November 1, 2021); 2.7 Genes Reveal How Some Rockfish Live up to 200 Years by Jack Tamisiea (November 11, 2021); 2.8 Here's How a Python Jaw Can Fit a Whole Deer by Bethany Brookshire (December 7, 2022); 3.1 Deepest Fish Discovered More Than 5 Miles below the Sea Surface by Tom Metcalfe (April 7, 2023); 3.2 Surprise Fossil-Munching Sponges Found in Arctic Seafloor Wasteland by Jack Tamisiea (February 8, 2022); 3.3 These Sharks Hold Their 'Breath' to Stay Warm by Meghan Bartels (May 11, 2023); 3.4 How the Woolly Bear Caterpillar Turns into a Popsicle to Survive the Winter by Kate Furby (March 3, 2023); 3.5 No One Knows How the Biggest Animals on Earth—Baleen Whales—Find Their Food by Kate Wong (April 1, 2023); 4.1 Why Has a Group of Orcas Suddenly Started Attacking Boats? by Stephanie Pappas (May 24, 2023); 4.2 Gorillas' Resilience after Early-Life Trauma Holds Lessons for Humans by Rachel Nuwer (May 23, 2023); 4.3 Parrot Babies Babble Just like Us by Karen Hopkin (May 12, 2023); 4.4 Bumblebees Show Off Their Own Puzzle-Solving Culture by Rachel Nuwer (March 7, 2023); 4.5 What Scientists Are Learning about Women's Health from Other Female Animals by Barbara Natterson-Horowitz (March 1, 2023); 4.6 Humans Are Not the Only Creatures Who Mourn by Barbara J. King (May 1, 2017); 4.7 The Social Genius of Animals by Katherine Harmon Courage (January 1, 2015); 4.8 Insects Recognize Faces Using Processing Mechanism Similar to That of Humans by Elizabeth A. Tibbetts, Adrian G. Dyer (December 1, 2013); 4.9 A Famed Dolphin-Human Fishing Partnership Is in Danger of Disappearing by Rachel Nuwer (January 30, 2023); 5.1 Two-Headed Worms Tell Us Something Fascinating about Evolution by Ashleigh Papp (May 10, 2022); 5.2 Killer Whales Are Speciating Right in Front of Us by Rüdiger Riesch (November 1, 2016); 5.3 'Chatty Turtles' Flip the Script on the Evolutionary Origins of Vocalization in Animals by Pakinam Amer (December 2, 2022)

Each author biography was accurate at the time the article was originally published.

INDEX